Fundamental Theories of Physics

The international monograph series "Fundamental Theories of Physics" started with a new Editorial Board from volume number 172 on. Earlier volumes which appeared in the names of the new Editorial Board were still reviewed and published under the responsibility of the previous Editors of the series.

Volume 174

For further volumes:
http://www.springer.com/series/6001

The series "Fundamental Theories of Physics" aims at stretching the boundaries of mainstream physics by clarifying and developing the theoretical and conceptual framework of physics and by applying it to a wide range of interdisciplinary scientific fields. Original contributions in well-established fields such as Quantum Physics, Relativity Theory, Cosmology, Quantum Field Theory, Statistical Mechanics and Nonlinear Dynamics are welcome. The series also gives a forum to non-conventional approaches to these fields. Publications should provide perspectives and carefully bridge conventional views with the presented new and promising ideas.

Although the aim of this series is to go beyond mainstream physics, a high profile and open-minded Editorial Board will carefully select the contributions and will ensure the high scientific standard of this series.

Peter Mittelstaedt

Rational Reconstructions of Modern Physics

Second Enlarged Edition

Edited by Kristina Engelhard

Springer

Peter Mittelstaedt
Institut für Theoretische Physik
Universität zu Köln
Köln, Germany

ISBN 978-94-007-9570-9 ISBN 978-94-007-5593-2 (eBook)
DOI 10.1007/978-94-007-5593-2
Springer Dordrecht Heidelberg New York London

Printed on acid-free paper

Springer is part of Springer Science+Business Media (www.springer.com)

Preface

The present book, rational reconstructions of modern physics, has evolved from investigations, lectures, and discussions with many colleagues in physics and philosophy during the last 10 years. Selected problems of this treatise were presented at various conferences, as the biennial meetings of the "International Quantum Structures Association" (IQSA) in 2002, 2006, and 2010, and at the annual conferences of the "Académie Internationale de Philosophie des Sciences" (AIPS), for instance in 2004. In particular, I mention here the lectures and discussions that I could contribute to the informal research seminars in philosophy of physics, which were organized by Brigitte Falkenburg at the University of Dortmund over a period of several years. The stimulating discussions at all these events, the critique of my new approach but also the encouragement to continue this way of reasoning, are gratefully acknowledged.

The aim of this book is to summarise the results of these efforts which were partly scattered throughout various journals, proceedings of conferences, festschrift-volumes etc. and to reorganize them in a new and systematic order. The results and implications of the present investigations are partly new and they are in general not in accordance with the well known interpretation of modern physics in the light of classical physics. The goal of this attempt is a rational reconstruction of the leading theories of modern physics, the theories of special and general relativity and quantum mechanics, a project that will be further elucidated and motivated in the "Introduction" of the main text.

Preface – Of the Second Edition

The second edition of the present book differs from the first one by the incorporation of several new arguments and considerations that were not yet contained in the first edition. However, the main goal of the book – the rational reconstruction of the theories of modern physics and the discussion of its consequences – is not changed. In particular, the following changes were made in the second edition:

1. In Section 2.6 of the first edition general relativity was sketched only very briefly, since a rational reconstruction of Einstein's field equations turned out to be inaccessible. However, the most interesting aspects of general relativity, its old and new interpretation, are not restricted by this deficiency. For these reasons, in Chapter 2 (reconstruction) and Chapter 5 (interpretation) of the second edition the role of general relativity was reconsidered and elaborated in more detail, also with respect to cosmological aspects. (Cf. Sections 2.6.1, 2.6.2, 2.6.3, 5.3.1, and 5.3.2)
2. The statistics of physical objects is treated in Section 3.6. As already shown in the first edition, the statistics of indistinguishable objects can be obtained merely by rational reconstruction and without any recourse to quantum mechanics and to the statistics of bosons. This way of reasoning is extended in the present new edition to Fermi statistics and that again without referring to electrons, atoms, etc. as quantum objects, i.e. merely by applying the method of rational reconstruction (Cf. Sections 3.6.1 and 3.6.2).
3. Chapter 5 is concerned with the problem of interpretation of modern physics. In the second edition, large parts of this chapter were written completely new and the entire chapter represents now one of the most important parts of the book. We show that theories that can be obtained by rational reconstruction allow in general for a direct and literal interpretation of their results. In Chapter 5 this result is demonstrated for the theories of special relativity, general relativity and quantum mechanics.

Editor's Preface – Of the Second Enlarged Edition

For health reasons Peter Mittelstaedt asked me to take over editorial responsibility for the second enlarged edition of the *Rational Reconstructions of Modern Physics*. At the time I took over, the first proofs were being made. My role has been to carry out the proofreading, incorporate some changes of content that Peter Mittelstaedt wanted to make on the newly written Section 3.6 and supervise the process of production. I did not intervene in language or style. All changes that were made were finally authorised by the author.

Bonn
March 2013

Kristina Engelhard

Acknowledgements

In the first place, I wish to thank Brigitte Falkenburg for her continuous, stimulating advice over the last 10 years and for the encouragement to write this book. Furthermore, I thank Kristina Engelhard for the discussion of many detailed questions of my partly unconventional proposals. – The kind and very effective cooperation with Tobias Schwaibold of *Springer* was a pleasure and should be gratefully acknowledged.

For the preparation of the second edition, I want to thank in the first place Helmut Fink who read very carefully the first edition in search of small misprints in orthography and punctuation. Furthermore, I thank again Kristina Engelhard for the discussion of numerous detailed questions of the new version of the book. Finally, the kind and effective cooperation with Aldo Rampioni of *Springer* should be gratefully acknowledged.

Contents

Introduction

Even without a fully elaborated idea of the development of physical theories, we expect – at least among physicists – a wide spread agreement with the hypothesis that progress in physics consists of an interplay between experimental results and theoretical drafts. A successful theory summarises a large number of experimental results in a formal mathematical and conceptual system, a so-called theory, where new and additional experiments will contribute either to a confirmation or – what is usually more important – to a refutation of the theory by means of falsification. A refutation of this kind is then a challenge to formulate a new theoretical concept, an improvement of the first preliminary theory. In this way, the development and the progress of physics seems to consist in a stepwise accumulation of new results and thus in a permanent increase of knowledge.

The idea of the dynamical development of physical theories as it was conceived by Thomas Kuhn[1] and others can also be incorporated into this very general conception, however with an important additional distinction. Whereas in the phases of "normal science" the accumulation of physical knowledge takes place by summarising more and more experimental results into an already existing theory – whose domain of validity is extended in this way often by rather artificial assumptions – in the phases of "revolution" the extended and exhausted old theory is replaced by a completely new theory, which will again be subject to the interplay of confirmation and falsification. One clue of this argument is that physical theories cannot really be falsified since in most theories there are ways to extend the theory by additional assumptions such that by fully exhausting the new theory all known results can be incorporated. Hence, from time to time "revolutions" seem to be unavoidable.

Accordingly, the progress in physics seems to consist of an increase of knowledge, of the increasing number of experimental results and of permanently improved theories that summarise these results and interpret them on the basis of theoretical connections between various results. There is, however, the important

[1] Kuhn (1962) and Feyerabend (1970).

argument of the advocates of Kuhn's ideas, that the formulation of a new theory after a revolution, is by no means unique but depends, except from the scientific situation, also on the historical, sociological and psychological background of the scientists involved. We will not go into the details of the long lasting controversial debate about the justification of these arguments, since for the most important theories of modern physics, the theory of relativity and quantum theory, there are essentially no alternative approaches known.[2]

The present investigation will not follow these two ways of reasoning, neither the traditional idea that the progress in science consists of continuous accumulation of new results, nor Kuhn's modification of this idea that the development of physics takes place in a sequence of large steps, which correspond to phases of revolution and consolidation. Instead of these well known alternatives, we argue in favour of a completely different way for explaining the progress in physics, in particular of physical theories in the last century. We will show that the two models mentioned for the progress in science are not able to grasp the most radical and at the same time very simple change from the so-called classical Newtonian physics to the theories of "modern physics" in the twentieth century.

In particular we will show that the transition from classical physics to the three leading theories of modern physics, special relativity, general relativity and quantum mechanics cannot adequately be understood as an increase of knowledge about various new empirical facts. On the contrary, the very progress of these transitions consists of a stepwise reduction of prejudices, i.e. of quite general hypothetical assumptions of classical mechanics that can be traced back to the metaphysics of the seventeenth and eighteenth centuries. Accordingly, our proof of these statements will be a constructive one: We start from Newton's classical mechanics and show that by abandoning or relaxing the various unjustified metaphysical hypotheses contained in it, the theories of "modern physics" can be constructed. In this way, we can demonstrate two important results. On the one hand we show how the theories of modern physics can be justified without explicit reference to new experimental results. On the other hand, the original difficulties of understanding the new theories can now be explained convincingly and at the same time completely eliminated.

---

[2] Except perhaps for the "Bohm" theory of quantum mechanics, and the "Jordan-Brans-Dicke" theory of general relativity.

Chapter 1
Rise and Fall of Physical Theories

1.1 The Evolution of Modern Physics from the Classical World

Classical physics, which is assumed here as represented by classical mechanics, is usually considered as the correct physical theory of our ordinary everyday experience (*OE*). This experience refers to the dimensions and processes of the human life, to slow motions, to time intervals that are comparable with hours, days, and years etc. and to distances that correspond almost to the dimensions of human beings. The world as we know it through ordinary experience determines what we call *intuitive* or *comprehensible*. We do not want to begin by spelling out the meaning of the word "intuitive" and its history. Rather, this concept will find its bearing in ordinary experience (*OE*), which, as pre-scientific experience, precedes all scientific cognition. This vague sense of *intuitive* and *comprehensible* is not only used in popular science but also corresponds largely to the usage of the terms in the literature of modern physics. This does not rule out the possibility that individual physicists interested in fundamental questions have understood these concepts in a deeper sense and in line with the philosophical tradition.[1] Such an interpretation, however, seems initially to have been limited to quantum mechanics and will therefore be omitted for the present purpose.

At this point, it is already useful to introduce an important distinction. The concept of intuitiveness that has its bearing in ordinary experience (*OE*) can be understood in two different ways.[2] First, it can be taken to mean the immediately and directly discernible intuitiveness, i.e. the agreement of a new experience with already known and familiar elements of ordinary experience. In this case, we also speak of *direct* intuitiveness. "Intuitive", however, can also refer to a familiarity or

[1] Falkenburg (2006).

[2] Huber (2006).

P. Mittelstaedt, *Rational Reconstructions of Modern Physics*, Fundamental Theories of Physics 174, DOI 10.1007/978-94-007-5593-2_1,

resemblance with ordinary experience, which is revealed to the observer only after prolonged observation and via a number of logical steps. In this case, we speak of *indirect* intuition. Both types of intuition are found in physics.[3]

Hence, there exists a kind of reductionism of intuitiveness. By means of a more or less long series of steps, phenomena that at first glance are by no means *directly* intuitive, comprehensible and familiar, can nevertheless be traced back to *directly* intuitive phenomena. These are the *indirectly* intuitive processes and experiential structures. Obviously, the indirectness of intuitiveness can differ by degree with respect to various phenomena and different observers. There are experiences, particularly in modern physics, the *indirect* intuitiveness of which has only been gradually recognized over the span of several decades, while simple, although initially irritating classical-mechanical processes – such as the Coriolis forces in an accelerated frame of reference – can be traced back very quickly to *directly* intuitive structures. Between different observers there can also be gradual differences. A trained and experienced natural scientist will be able to discern the *indirect* intuitiveness of a process more quickly and in fewer steps than a lay person in the same area.

Classical physics sees itself as the scientific theory of the world of ordinary experience. However, in the formulation of classical physics, especially of classical mechanics (*CM*), certain basic experiences are extrapolated beyond immediate experience and elevated to general principles, which then form the basis of the corresponding classical-physical theory. These principles, which are not justified by experience alone, can probably be attributed to the influence of modern metaphysics in the rise of classical physics in the seventeenth and eighteenth centuries.[4] This influence concerns the assumption of the existence of an absolute space, the assumption that Euclidean geometry applies to this space, the existence of an absolute and universal time as well as the assumption of an unbroken causality. Furthermore, there is the assumption that matter consists of individual and permanent substances, the properties of which are "thoroughgoingly determined" and simultaneously decidable.

We summarise these hypothetical assumptions in the "classical ontology" $O(C)$. Generally, by the term "ontology" we understand the most general features of a certain domain of reality. In the present case, we call "classical ontology" $O(C)$ the most general properties of objects that belong to the realm of classical physics, i.e. to the classical world. In a similar sense, we will later speak of "quantum ontology" $O(Q)$ referring to all kinds of quantum objects. In the following, we often compare ontologies with respect to their strength. Generally, we say that an ontology O is stronger than another ontology O', if an object o contained in O must fulfil more requirements than an entity o' contained in O'. By these explanations we establish a partial ordering relation between two ontologies O and O'.

---

[3] Huber (2006) speaks of "sensible intuitive" and "rationally intuitive".

[4] Falkenburg (2006).

Ordinary experience (*OE*) does not justify the ontological hypotheses of the classical ontology *O*(*C*). But ordinary experience is also not precise enough to contradict them. This means, that in any case classical mechanics is loaded with ontological hypotheses that have neither a rational nor an empirical justification. Hence, since these presuppositions enter explicitly into classical mechanics, it is to be expected, that classical mechanics will enter with physical experience as that experience becomes more and more precise. In order for us to be able to respond adequately to such contradictions, we shall now investigate in more detail where these ontological hypotheses enter as premises into the structure of classical mechanics. Once that is known, the hypotheses could possibly be weakened or abandoned. This weakening corresponds to a transition from the classical ontology *O*(*C*) to a weaker ontology *O*' in the above mentioned sense.

Since the formulation of classical mechanics in Newton's *Principia* of (1687)[5] neither classical mechanics itself nor the underlying classical ontology *O*(*C*) were seriously questioned by physicists or by philosophers for almost 200 years. Only at the end of the nineteenth century, Ernst Mach[6] and Henri Poincaré[7] formulated serious objections against the conceptual basis of classical mechanics. A few years later, the same arguments and objections were taken up by Albert Einstein for establishing the special theory of relativity. However, Einstein did not discuss explicitly the arguments of Mach and Poincaré against classical mechanics and Newton's theory of space-time. Moreover, it is an open historical question whether Einstein knew at all the objections of Mach and Poincaré in detail from the literature. With respect to Ernst Mach, Stachel[8] mentions the possible influence of Einstein's early reading of Mach's *Mechanics* (around 1897). But nothing is known about Einstein's reaction to the work of Poincaré. Presumably, he never read Poincaré's most relevant paper of 1898. In addition, Einstein and Poincaré never met before the first Solvay Conference in 1911, where they had only a short and ineffective debate.[9]

In detail, the critique of Mach and Poincaré was concerned with the concepts of absolute space and absolute time in Newton's *Principia*. With respect to time Newton wrote:

> Absolute, true and mathematical time, of itself, and from its own nature flows equably without regard to anything external, and by another name is called duration: relative, apparent and common time, is some sensible and external (whether accurate or unequable) measure of duration by the means of motion, which is commonly used instead of true time.[10]

Mach emphasised that this concept of *absolute time* has no empirical meaning, since there is no realisable experiment that could be used for measuring numerically

---

[5] Newton (1687).

[6] Mach (1901).

[7] Poincaré (1898).

[8] Stachel (1989).

[9] Huber (2000).

[10] Newton (1934), p. 6.

the time difference between two arbitrary events. In addition, he could show that there are no motions that can be used as reliable and universal clocks. Which kind of motion we consider to be a clock is merely a matter of convention. In summarising these results, Mach argued that *absolute time* is a useless metaphysical concept and should be completely eliminated in physics. On the basis of similar considerations, Henri Poincaré arrived at the result that the measure of time that we use in physics, is not based on empirical grounds but on arbitrary conventions and on non-empirical principles like simplicity. Another aspect of time, which was first clarified by Poincaré, is the concept of distant simultaneity. Using an astrophysical example, Poincaré could show that any concept of simultaneity of events separated in space is based on conventional stipulations that are – in most cases – tacitly presupposed in our physical theories.

A similar, but not completely equivalent problem is induced by the concept of absolute space. In the *Principia* Newton wrote:

> Absolute space, in its own nature, without regard to anything external, remains always similar and immovable. Relative space is some movable dimension or measure of the absolute spaces; which our senses determine by its position to bodies: and which is vulgarly taken for immovable space.[11]

In contrast to the problem of absolute time, Newton was convinced that the *absolute space* has a well defined empirical meaning and he described explicitly an experiment – the so-called bucket experiment – that would allow one to decide whether a given body is at rest in absolute space or whether it is moving relative to the absolute space. In this way, the concept of absolute motion could be defined. In this situation, Mach could not simply argue that in the same sense as absolute time also absolute space is a purely metaphysical concept, that should be eliminated in physics. Hence, Mach demonstrated in a first step, that Newton's interpretation of the bucket experiment and thus the experimental proof for the existence of an absolute space was a fallacy – and on the basis of this demonstration he argued that also the absolute space is a metaphysical concept that should be eliminated in physics.

Similar arguments against classical mechanics were put forward by the founders of quantum mechanics (*QM*) many decades later. Bohr, Heisenberg, Schrödinger and others found out that several most general features of classical mechanical objects were no longer in accordance with theoretical and experimental results of the new quantum physics. We mention here the thoroughgoing determination and individuality of objects, the unbroken strict causality of physical processes and the general conservation of substance. Obviously, these results confirmed the above mentioned observations made in special relativity, that Newton's classical mechanics and space-time theory is loaded with metaphysical hypotheses without any rational or empirical justification.

Summarizing the results of this section (a) we find that classical physics is loaded with hypotheses that cannot be confirmed or disproved in the realm of ordinary experience (*OE*). Hence, (*OE*) is not adequately described by classical

[11] Newton (1934), p. 6.

mechanics. In modern physics, i.e. in the physics of the twentieth century, which investigates the properties of space-time and of the quantum world with a much higher degree of accuracy than in earlier centuries, the hypotheses mentioned were even found to be in contradiction with theoretical and experimental results of special relativity (*SR*), general relativity (*GR*), and quantum mechanics (*QM*), respectively. On the basis of these results, we can elucidate also another interesting aspect of the interrelations between classical physics – represented by classical mechanics – and modern physics given by (*SR*), (*GR*), and (*QM*), namely the problem of intuitiveness.

1.2 Intuitiveness and Truth of Modern Physics

Since its emergence, twentieth century physics has had the reputation of being unintuitive, abstract and difficult to understand, while the earlier classical physics, by contrast, is regarded as intuitive and comprehensible. The history of physics in the twentieth century demonstrates how this widespread assessment resulted in a clear rejection of the new theories, the defamation of individual scientists and even political persecution. Einstein's special theory of relativity[12] of 1905 was especially affected by this rejection, probably also because it was the first of the theories of modern physics to be apparently in clear contradiction to "common sense". The, in many respects, much more radical general theory of relativity of 1916, on the other hand, provoked much less public agitation and rejection, a fact that was probably due also to its complex mathematical form. The continued interest in the cosmological consequences of this theory such as the big bang, the age of the universe, cosmic expansion etc. is due to the significance of these results for our world view and not to the alleged unintuitiveness of the underlying theory. It was not until quantum mechanics was discovered in 1925 that there was again considerable public irritation due to the numerous philosophically explosive theses contained in this theory.

In the following reflections, we shall consider whether there is a *factual* justification for the above mentioned assessment of classical and modern physics, which is based merely on a historical observation. Initially, this impression seems to be confirmed by the fact that modern physics, i.e. the theory of relativity and quantum mechanics, deals with previously unknown phenomena that do not occur in our ordinary experience (*OE*). The difficulties with the acceptance of modern physics, however, could also be due to the fact that the intuitiveness of its phenomena is not easily recognized and not at first glance. It will turn out that this is actually the case. According to the definition of the concept of intuitiveness by recourse to ordinary experience (*OE*) we could try to answer the question whether in this sense classical physics is intuitive and modern physics unintuitive. The first impression seems to

[12] Hentschel (1990), Huber (2000), and Könneker (2001).

support these two theses. There is, however, still another important argument which must be taken into account. It concerns the observation mentioned above, that classical physics is loaded with metaphysical hypotheses, which cannot be justified by rational or empirical means – and which do not occur any longer in the theories of modern physics.

The following considerations will show, that these two arguments – the problem of intuitiveness in classical physics and modern physics, and the role of metaphysical hypotheses in physical theories – lead to an adequate understanding of the peculiar relation between classical and modern physics. In particular, these reflections will show how the theories of modern physics could evolve from theories of the classical world. The objective of the present study is systematic rather than historical. We are not interested in knowing how the great three theories of modern physics – the special theory of relativity, the general theory of relativity, and quantum mechanics – actually came about and what intentions their discoverers pursued. Neither do we want to focus on the reception of these theories. Instead, we want to discuss the purely theoretical question how these three theories could have been discovered if, step by step, scientists had freed themselves from metaphysically motivated hypotheses contained in classical mechanics. The answer to this question yields surprising results regarding the intuitiveness and truth of modern physics.

1.3 The New Approach: Reduction and Elimination of Metaphysical Hypotheses

On the basis of the preceding discussion it became obvious that classical mechanics (*CM*) and, more generally, the classical ontology $O(C)$ is loaded with hypotheses that are neither plausible nor intuitive. Since the theories of modern physics seem to be free in each case from one ore more of these hypotheses, it could perhaps be possible to obtain essential features of these modern theories by relaxing or even eliminating some of the hypothetical assumptions of the ontology $O(C)$ of classical mechanics. This is an interesting task and we will follow this idea in the subsequent chapters. We will try to answer the question whether simply eliminating or weakening the hypotheses mentioned already leads to the theories of modern physics, or whether other assumptions, empirical results, or new hypotheses must be added.

We do not maintain that the components $O(C)^1 \ldots O(C)^6$ of the classical ontology listed below are complete. We mention here only those ontological features that are known to be in disagreement with one or more of the theories of modern physics. Hence, we must be prepared to find out one day more, not yet discovered ontological properties, that provide similar but new problems with future discoveries in physics. – In addition, the elimination of metaphysical assumptions in physical theories must not be completely exhaustive. A rigorous abandonment of all metaphysical assumptions in physics would possibly destroy indispensable conceptual prerequisites of any physical theory, as the possibility of a formal language of physics, the assumption of a world that exists irrespective of our consciousness, and the existence of a universal intersubjective natural science that is independent of individual scientists etc.[13]

Even if it is obvious that a reduction of the ontological hypotheses of classical mechanics can be carried through in the described way – there are still some open questions. Is it really sufficient to abandon completely just those ontological hypotheses that we know accidentally, or could it happen that in some cases we would thereby go too far? In this case, the reduction itself must be weakened. The reduction could then be performed in two steps. First, we eliminate a strong hypothesis and secondly we add a much weaker hypothesis as a compensation. This is not only a hypothetical problem. In the following considerations we will find out that a situation of this kind actually does occur in the realm of Quantum Mechanics, which is treated in Chapter 3 in more detail.

In order to carry through a reconstruction of one of the great three theories of the twentieth century physics, we begin with a detailed formulation of the classical ontology $O(C)$.

The ontological hypotheses of classical mechanics read:

$O(C)^1$ There exists an absolute time. It establishes a universal temporal order of two or more events, it provides a universal measure of time and it explains the concept of simultaneity of two spatially separated events.

$O(C)^2$ There exists an absolute space. It explains the concepts of absolute motion and absolute rest. Euclidean geometry applies to this absolute space.

$O(C)^3$ There are individual and distinguishable objects. These objects cannot only be named and identified at a certain instant of time, but also re-identified at any later time. Generally, they possess also a temporal identity.

$O(C)^4$ These objects possess elementary properties P_λ in the following sense: An elementary property P_λ refers to an object system such that either P_λ or the counter property $\bar{P}_\lambda$ pertains to the system. Furthermore, objects are subject to the law of thoroughgoing determination according to which “if all predicates are taken together with their contradictory opposites then one of each pair of contradictory opposites must belong to it”.[14]

$O(C)^5$ For objects of the external objective reality, the causality law holds without any restriction. There is an unbroken causality.

$O(C)^6$ For objects of the external objective reality the law of conservation of substance holds without any restriction.

The classical ontology $O(C)$, which is characterised by these requirements, is neither intuitive and comprehensible nor is it justified by experimental evidence.

[13] Vollmer (2000, pp. 46–67, 2007, pp. 67–81).

[14] Kant (1998, B599).

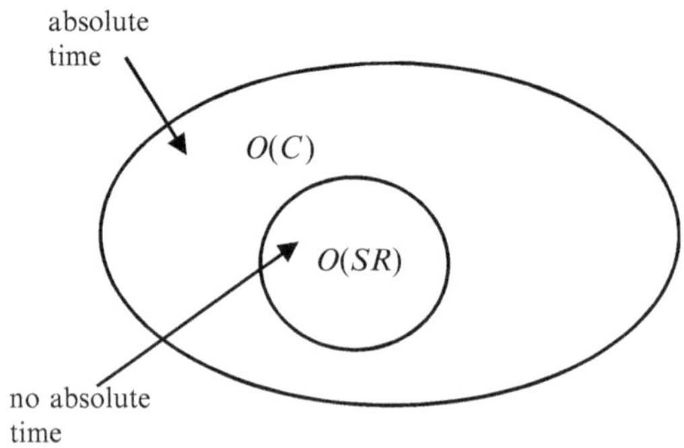

Fig. 1.1 The step from the classical ontology $O(C)$ to the improved ontology $O(SR)$ of special relativity

It is overloaded with metaphysical hypotheses that clearly exceed our ordinary experience (OE), which on its part determines what we call *intuitive* and *comprehensible*. Hence, classical mechanics, which is based on classical ontology is not the physical theory of our ordinary everyday experience. In addition, it is also not the correct theory of modern physics, classical mechanics describes a fictitious world which does not exist in reality.

In the subsequent Chapters 2 and 3 we will be presented with three different kinds of reconstructions by relaxing and eliminating metaphysical hypotheses of classical mechanics. These different kinds correspond to the various problems of eliminating the hypothetical assumptions just mentioned.

(a) In Chapter 2 we present a reconstruction of special relativity by abandoning the hypothesis $O(C)^1$, which states the existence of absolute time. In this case, our way of reasoning can be applied perfectly to the problem. In a first step we demonstrate how some general structures of space and time can be established and how some basic concepts of mechanics can be formulated. In this part, it does not matter whether we make use of the hypothesis $O(C)^1$ of the existence of an absolute time. However, in a second step of our derivation it matters whether we make use of this hypothesis or not. If we use the hypothesis $O(C)^1$ we obtain Newton's classical space-time with Galilei transformations and the basic elements of classical mechanics. If we abandon completely all assumptions that can be traced back to the hypothesis $O(C)^1$, then we arrive at the Minkowskian space-time of special relativity, in particular at the Lorentz-transformations – and that without adding any new assumption – neither a theoretical one (an additional hypothesis) nor a new empirical result (Fig. 1.1).
(b) The attempt to reconstruct in a similar way also the general theory of relativity (in Section 2.6) is less convincing. The reason is that we are confronted here with the situation that the elimination of classical ontological hypotheses – $O(C)^1$ and $O(C)^2$ – is not sufficient. The abandonment of $O(C)^1$ leads merely to Special Relativity, as will be shown later in Chapter 2.2. The hypothesis $O(C)^2$ consists of two parts, the assumption of an absolute space that defines the concept of absolute motion, and the assumption that Euclidean geometry applies to this space. Since Newton's assumption of the existence of an absolute

Fig. 1.2 The steps from classical ontology $O(C)$ to the reduced ontology $O(GR)$ of general relativity

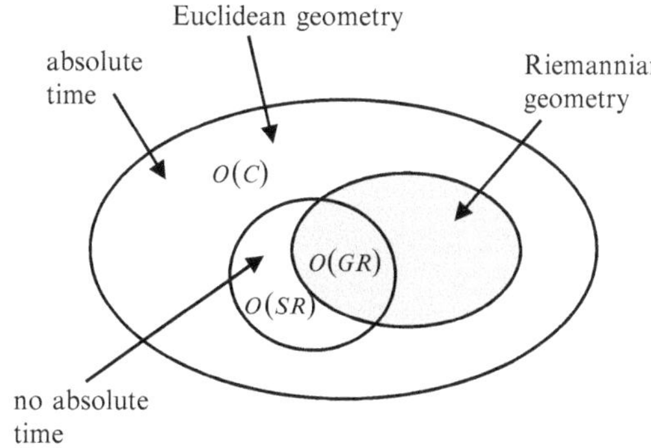

space is completely absent in the foundations of classical mechanics – which was not recognized by Newton – there is no need to eliminate this hypothesis. However, the assumption that Euclidean geometry applies at least to the "relative space", is important for the mathematical formulation of classical mechanics. There is, however, no need to eliminate this assumption completely. It must merely be relaxed. In this way we could arrive at a Riemannian geometry of the 3-dimensional space and at a pseudo-Riemannian Geometry (with signature 2) of the 4-dimensional space-time continuum. In spite of this encouraging result, this is not yet the entire general relativity but only the geometrical part of it. The missing part consists of Einstein's field equation and the coupling between space-time and gravitation, which is induced by these equations. The coupling between matter, the source of the gravitational field, and the geometry of space-time, cannot be obtained merely by eliminating or relaxing convenient hypotheses $O(C)^k$ of classical mechanics or classical ontology, respectively (Fig. 1.2).

(c) In Chapter 3 we describe the attempt to reconstruct quantum mechanics, or more precisely non-relativistic quantum mechanics in Hilbert space. Since the relativistic aspect is not relevant here, we leave the hypotheses $O(C)^1$ and $O(C)^2$ unchanged and relax only the remaining hypotheses $O(C)^3 \ldots O(C)^6$. However, for the relaxation of these ontological assumptions we must apply a somewhat modified strategy. In particular, the hypothesis $O(C)^4$ turns out to be too global and not sufficiently differentiated. For this reason, a complete abandonment of this hypothesis would destroy too much and not lead to the desired result. Hence we must either compensate the complete abandonment by adding a new, weaker hypothesis, or we should abandon merely an ontological hypothesis $O(C)^{4*}$ that is definitely weaker than $O(C)^4$. In both ways we arrive at a formulation of quantum mechanics that contains not only the idea of mutually excluding complementary observables, but also its relaxation of joint unsharp, statistically complementary properties.

We will explain this problem more in detail later in Sections 3.5 and 4.3. For the present it is sufficient to illustrate our strategy in Fig. 1.3. We have to perform two tasks: First we must reduce classical ontology $O(C)$ by eliminating

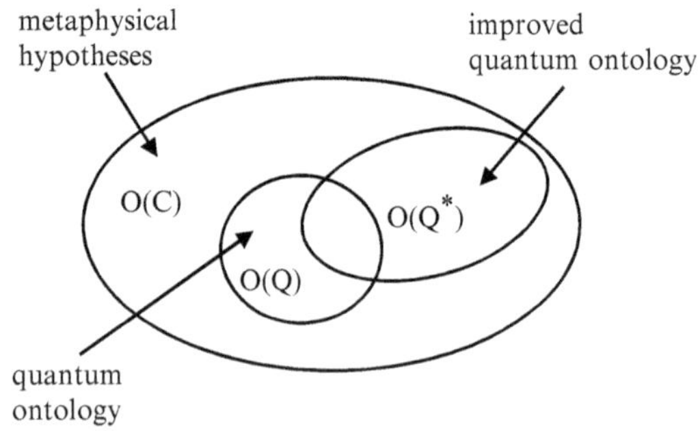

Fig. 1.3 Two steps from classical ontology $O(C)$ to the improved quantum ontology $O(Q^*)$

not sufficiently justified hypotheses. This corresponds to the transition $O(C) \rightarrow O(Q)$ to a quantum ontology $O(Q)$. In a second step we must correct partly the first step, since the ontology $O(Q)$ is both too restrictive and also not sufficiently restrictive. This step is indicated by the transition $O(Q) \rightarrow O(Q^*)$ to a more sophisticated ontology $O(Q^*)$.

Chapter 2
Reconstruction of Special and General Relativity

2.1 Historical Development Versus Rational Reconstruction

The historical development of the theory of special relativity offers a rather complicated and confusing impression. At the end of the nineteenth century we find several important philosophical investigations by Ernst Mach and Henri Poincaré about the underlying philosophical prejudices of Newton's theory of space and time and of classical mechanics. In addition, we find important mathematical contributions by Poincaré and Lorentz about the structure of space and time. Finally, there was an extensive discussion about the meaning of the Michelson experiment, which was considered – erroneously – by many physicists as an *experimentum crucis* for the validity of special relativity. Actually, the Michelson experiment demonstrates merely the isotropy of the so-called two-ways velocity of light. We will come back to this point later.

Without any explicit reference to these various considerations Einstein formulated the theory of special relativity in his famous paper of 1905, which terminated the long lasting debate about space, time, and relativity – at least among experts in this field. As already mentioned above (in Section 1.a), it is not quite clear whether Einstein knew the relevant contributions of Mach and Poincaré. Here we refer again to the investigations of Huber[1] and Stachel.[2] In addition, it is not known whether Einstein knew the Michelson experiment[3] in 1905,when he wrote his paper. Presumably, Einstein mentioned this experiment in his later publications merely as an *ex post* justification of his theory.

Under these conditions, it is not surprising that in 1905 Einstein did not provide a theory, that distinguishes clearly between empirical results and hypothetical elements in the foundation of his publication. Presumably, also the difficulties of the acceptance of the new "theory of relativity", as it was called very soon, can

[1] Huber (2000).

[2] Stachel (1989).

[3] Stachel (1982).

P. Mittelstaedt, *Rational Reconstructions of Modern Physics*,
Fundamental Theories of Physics 174, DOI 10.1007/978-94-007-5593-2_2,

at least partly be explained by this lack of systematics. Accordingly, after a few years of consolidation of the new theory, several physicists tried to reformulate and to simplify the theory of special relativity in terms of an axiomatic theory making use of the "principle of relativity" and the "principle of the constancy of the velocity of light" as basic requirements of the theory.

We will not follow here this strategy and this way of reasoning, but according to our general program we will establish a "rational reconstruction" of special relativity which does not refer to the historical development of the theory at all. Instead, we want to discuss the pseudo-historical question, how special relativity could have been discovered if scientists had freed themselves from the metaphysically motivated hypotheses contained in Newton's theory of space-time and classical mechanics and of other widespread prejudices like the misleading interpretation of the Michelson experiment.

The reconstruction of special relativity that we have in mind here has a long history and can be traced back to the early days of (*SR*) in the beginning of the twentieth century,[4] but at this time it was almost ignored by the scientific community of physicists. Of course, in the contributions quoted here, the authors were not able to identify an invariant velocity constant contained in the results of this approach with the velocity of light. In addition, this way of reasoning was motivated by the somewhat antiquated and no longer convincing idea that special relativity should be based on two axioms, the postulate of relativity and the postulate of the invariance of the speed of light.[5] Many years later, this attempt was elaborated in more detail[6] and combined with the idea of a rational reconstruction of a physical theory that is free from metaphysical prejudices.[7] The present paper follows essentially the way of reasoning of these latter publications.

2.2 Reconstruction of Special Relativity

The reconstruction of special relativity occurs in several clearly distinguished steps, which will briefly be mentioned here. The first step consists of the introduction of space and geometry. However, for two reasons we will not make use of Newton's hypothesis $O(C)^2$ of the existence of an absolute space, which we briefly discussed in Section 1.1. First, as we mentioned already in Section 1.1, the pretended experimental proof of this hypothesis by means of the "bucket-experiment" was shown by Ernst Mach[8] to be a fallacy. Second, for the foundations of classical mechanics, the assumption of the existence of an absolute space is not needed at all. Ironically,

[4] Ignatowski (1910) and Franck and Rothe (1911).

[5] We will not discuss here the reasons for the inadequacy of this axiomatic approach.

[6] Levy-Leblond (1976) and Mittelstaedt (1976).

[7] Mittelstaedt (1995, 2006).

[8] Mach (1901).

Newton's own mathematical formulation of classical mechanics does not make use of an absolute space and demonstrates in this way the redundancy of this concept. Instead of the strong and unnecessary hypothesis $O(C)^2$ we will make use of the much weaker assumption, that for any observer there is a local "relative" space that allows for the most general geometrical constructions. According to the first Helmholtz theorem, the assumption that finitely extended measurement rods are freely movable in space implies that the geometry measured with these measuring rods is elliptical, hyperbolical, or Euclidean.[9] This result makes it possible to establish an Euclidean space with the help of additional conventional postulates. Hence, we will presuppose here merely the free mobility of finitely extended measuring rods in a local relative space. We should, however, emphasise already at this place that the presupposition mentioned is realisable only if there are no gravitational fields. The reason is that gravitational fields cannot be screened off and thus there is no region in space that is free from gravitational forces, which is a necessary precondition for free mobility.

Our next task is the introduction of the concept of time. Since we have abandoned the hypothesis $O(C)^1$ of the existence of an absolute and universal time, for a constructive approach to the concept of time we proceed in the following way. Time is a one dimensional continuum which we describe by a coordinate θ. At any position x, time can be measured by a local clock $C(x)$ which is realized by a convenient physical process. This time scale θ, which measures merely the direction of time and the order of several events, will be called here *topological* time. In order to determine the *metric* time that is measured by a local clock, we proceed in three steps.

1. *Inertial system.* We consider an ensemble $\Gamma\{k_1, k_2, \ldots k_n\}$ of bodies freely thrust into space as well as a frame of reference equipped with measuring rods, which can be visualized as a material base of an observer (e. g. a spacecraft). If the trajectories of the test bodies k_i are Euclidean straight lines from the perspective of this frame of reference, then the frame of reference is called an inertial frame of reference or an *inertial system* denoted usually by I. It is obvious that this definition of inertial systems presupposes that there are no gravitational fields. Otherwise we would never find a frame of reference of the kind mentioned.
2. *Topological time.* In an inertial system I with spatial coordinates (x, y, z), at an arbitrary space point x a topological time θ can be introduced by means of an arbitrary physical motion. A topological time determines the direction of time and the temporal order of several events.
3. *Metric time.* While topological time can be tied to an arbitrary physical process, *metric* time is subject to the requirement that the test bodies k_i not only move on straight lines but also at constant velocities. This requirement can be met, since empirically the bodies k_i move uniformly relative to one another, which can be determined without knowledge of a metric time.

[9] More details about the two Helmholtz theorems can be found in Section 2.6.

In this way, at any spatial point x a metric time $t = t(x)$ is defined, which can be measured by a local clock $C(x)$. We restrict our considerations to one spatial coordinate and write $t(x)$ and $C(x)$. On account of its definition, metric time t is determined only up to a transformation

$$t \to t' = \alpha t + \beta, \quad \text{with constant numbers } \alpha \text{ and } \beta.$$

The constant α changes the temporal measure and constant β the zero point of the time scale. Both parameters must be determined by a convention.

The combination of space and time in a space-time continuum occurs through the definition of the synchronicity of spatially separated clocks and through the determination of the transformations between different inertial systems. The free choice of the zero-point of the metric time implies that for two clocks $C(x_A)$ and $C(x_B)$ at different points x_A and x_B in space their synchronicity must be determined by a convention.[10] Thereby, Einstein's method by means of light signals turned out to be particularly simple and useful. At time ${t_A}^{(1)}$ a light signal is emitted in x_A, which is reflected in x_B and received again in x_A at time ${t_A}^{(3)}$. In order to find out the zero-point of a clock $C(x_B)$ – of the same construction – in x_B, we determine the time ${t_B}^{(2)}$ of the reflection by the condition

$${t_B}^{(2)} = {t_A}^{(1)} + \varepsilon\left({t_A}^{(3)} - {t_A}^{(1)}\right)$$

where ε is an undetermined real parameter $0 \leq \varepsilon \leq 1$, that must be determined by a convention. In physics, usually Einstein-synchronisation is applied, where $\varepsilon = ½$. This convention implies that the velocity of light in the two possible directions has the same value. For methodological reasons we emphasise, that this convention does not imply that the metric of space-time depends essentially on the existence of light, which – in this case – would have an important constituting influence on the structure of space-time. The same convention could also be achieved by other methods, e.g. by slow motion of clocks.[11]

On the basis of these definitions, in an inertial system I we can introduce a system K_I of coordinates with three Cartesian coordinates (x, y, z) of the Euclidean space and a metric time t such that time scales of different positions in space are connected according to Einstein-synchronisation with $\varepsilon = ½$. We denote a system of coordinates $K_I\ (x, y, z, t)$ of this kind as a Galilean system of coordinates in the inertial system I. In a given inertial frame of reference I a Galilean system of coordinates $K_I\ (x, y, z, t)$ is not yet uniquely determined. Indeed, it is possible to change the system K_I of coordinates by several space-time transformations, without thereby changing the state of motion of the inertial system I. These internal transformations of an inertial system consist of:

---

[10] Poincaré (1898).

[11] Mittelstaedt (1976/89).

1. Translations in space

$$x_k \to x'_k = x_k + a_k, \quad a_k = \text{const} \quad (k = 1, 2, 3)$$

with a constant vector $a = (a_1, a_2, a_3)$, which shifts the spatial zero-point of K_I.
2. Orthogonal rotations in space

$$x_k \to x'_k = a_{ki} \cdot x_i \quad a_{ki} = \text{const.} \quad (i, k = 1, 2, 3)$$

with the orthogonality relations

$$a_{ik} \cdot a_{lk} = \delta_{il} = a_{ki} \cdot a_{kl}$$

induce a rotation of the spatial coordinate axis, where distances are conserved. Since the nine coefficients of a_{ik} must satisfy six relations, there are three independent parameters α_i.
3. Translations in time

$$t \to t' = t + \beta \quad \beta = \text{const.}$$

don't change the inertial system. The metric time is merely determined up to a transformation $t \to t' = \alpha t + \beta$, where parameter α determines the time-scale, which is not changed here, and β induces a shift of the zero-point of the time scale.

Hence we find, that the internal transformations of an inertial I system contain 7 parameters, 3 parameters a_i of the translation, 3 parameters α_i of the orthogonal rotation, and 1 parameter β of the shift of the zero-point of the time-scale.

2.2.1 Transformations Between Inertial Systems

For investigating transformations $T_{II'}$ between two inertial systems, we consider two inertial systems I and I' with relative motion that are oriented at the same constituting ensemble $\Gamma = \Gamma\{k_1, k_2, \ldots\}$. We will assume here that the systems of coordinates $K_I\,(x, y, z, t)$ and $K_{I'}(x', y', z', t')$, that belong to I and I', respectively, are Galilean systems of coordinates, such that t and t' are metric time scales, and the synchronisation parameter of systems I and I' reads $\varepsilon = ½$. For the sake of simplicity, we restrict our considerations to one spatial coordinate x in the following.

The transformations $T_{II'}$

$$x \to x' = f_i(x,\, t)$$

$$t \to t' = f_0(x,\, t)$$

that connect the systems of coordinates K_I (x, y, z, t) and $K_{I'}(x', y', z', t')$ can be derived from a few postulates that follow from the fact that both systems I and I' are oriented at the same constituting ensemble Γ. In system I as well as in system I' point like test bodies k_i are moving equably on straight lines. Hence we require

Postulate 1: Straight (x, t) lines are transformed into straight (x', t') lines.

According to the fundamental theorem of projective geometry, this postulate leads to the functions

$$x' = (a_1 x + b_1 t + c_1)/(ax + bt + c);$$

$$t' = (a_0 x + b_0 t + c_0)/(ax + bt + c)$$

with the same denominator in both fractions. In order to avoid unnecessary singularities, we require for further simplification.

Postulate 2: Finite values (x, t) *are transformed into finite values* (x', t').

This postulate implies the linearity of the functions f_0 and f_I,

$$x' = a_1 x + b_1 t + c_1, \quad t' = a_0 x + b_0 t + c_0.$$

Consequently, a point P' at rest in I' has in I the velocity

$$v_{P'} = (dx/dt)_{dx'=0} = -b_1/a_1 = \text{const.}$$

which means that system I' is moving relative to system I with a constant velocity $v_{II'} = v$. Hence the relative motion of two inertial systems is equably in time and straight in space. The coefficients a_i, b_i, *and* c_i of the transformation are functions of the relative velocity v.

For further simplification we require

Postulate 3: The systems of coordinates K_I (x, t) and $K_{I'}(x', t')$ coincide at $t = 0$.

This requirement implies $c_1 = 0$ and thus the spatial part of the transformation reads

$$x' = a_1(v)x + b_1(v)t = a_1(v)(x + b_1(v)/a_1(v) \cdot t)$$

and thus

$$x' = k(v)(x - vt)$$

where we used $k(v) := a_1(v)$ Hence, the formulas for the transformation read

$$x' = k(v)(x - vt), \quad t' = \mu(v)t + \nu(v)x$$

with three arbitrary functions $k(v)$, $\mu(v)$, and $\nu(v)$ of the relative velocity v.

2.2.2 Digression: Derivation of the Galilei Transformation

At this point of our derivation it becomes obvious that the assumption of an absolute and universal time t, as it is presupposed in Newton's theory of space and time, leads to the Galilei transformations. Indeed, if we use this universal time in the systems I and I', then we have $t = t'$ and thus $\mu(v) = 1$ and $v(v) = 0$ and obtain $x\prime = k(v)(x - -vt)$, $t' = t$ with only one unknown function $k(v)$. This function can be determined to be $k = 1$ in the following way.

Since the constituting ensemble Γ does not distinguish any particular direction in space, we require:

Postulate 4: *Isotropy* of the transformation $T_{II'}$

If we consider – as always in the present investigation – only one spatial coordinate x, then the requirement of *isotropy* means, that a simultaneous change of signs of x, x', and v does not change the transformation. Hence, the original transformation

$$x' = k(v)(x - vt), \qquad t' = t$$

is changed by the transition

$$(x, x', v) \to (-x, -x', -v)$$

into the new transformation

$$x' = k(-v)(x - v \cdot t) \quad t' = t$$

which must not differ from the original transformation. Comparing the coefficients of both transformations, we find

$$k(v) = k(-v)$$

i.e. $k(v)$ is an even function.

The constituting ensemble Γ does not distinguish one particular system of inertia. With respect to their defining properties, inertial systems are indistinguishable. We summarize this statement in the following

Postulate 5: No system of inertia is distinguished with respect to the ensemble Γ $(k_1, k_2 \ldots k_n)$ (*principle of relativity*).

Remark: The kinematical indistinguishability of inertial systems is often called their "relativity". The "principle of relativity", which asserts this indistinguishability, is obviously not an empirical principle but follows from the definition of an inertial system and of the ensemble Γ. In particular, it is not a characteristic of special relativity but it holds in the same sense in the Newtonian theory of space-time. Hence, also for the derivation of the Galilei transformations it must be taken into account.

For the transformation, the indistinguishability of the inertial systems I and I' means that the systems are also indistinguishable with respect to the connecting transformations $T_{II'}(v)$ with $v = v_{II'}$ and $T_{I'I}$ (v') with $v' = v_{I'I}$. Hence, the transformations $T_{II'}(v)$ and $T_{I'I}$ (v') must have the same formal structure. The application of this property allows for further specification of the transformation.

Using the above mentioned form $x' = k(-v)(x - -v \cdot t)$, $t' = t$ of the transformation, a comparison of the two transformations

$$T_{II'}(v)\colon \quad x' = k(v)(x - v \cdot t), \quad t' = t$$

$$T_{I'I}(v')\colon \quad x = k'(v')(x' - v' \cdot t'), \quad t = t'$$

leads in a first step to $k(v) = k'(v)$. The complete evaluation of the principle of relativity leads to the requirement that the transformations

$$\boldsymbol{T}^{-1}{}_{II'}(v)\colon \quad x = x'/k(v) + v \cdot t', \quad t = t'$$

$$T_{I'I}(v')\colon \quad x = k(v')(x' - v't'), \quad t = t'$$

must agree. Comparing the coefficients with respect to x' we get $1/k(v) = k(v')$ and with respect to t'

$$v = -k(v')v' =: u(v').$$

On account of the principle of relativity we have also

$$v' = u(v) = -k(v)v.$$

Hence, for the function $u(v)$ we obtain the functional equation

$$v = u(u(v))$$

with the solutions $u(v) = \pm 1/v$ and $u(v) = \pm v$. If we require that $u(v)$ is defined for all values of the relative velocity v, then the solution $u(v) = \pm 1/v$ must drop out. In order to decide between the remaining solutions we note that $u = v$ leads to $k(v) = -1$ and that $u = -v$ leads to $k(v) = +1$. If we require that the transformation

$$x' = k(v)(x - v \cdot t), t' = t$$

emerges continuously from the identity $T_{II'}(0)$, then we have $x' = k(0)x = x$ and obtain $k = 1$. Hence, also $u = v$ is dropped out and we obtain

$$u(v) = -v,\ v' = -v,\ k(v) = 1.$$

Finally, the transformation assumes the form

$$x' = x - v \cdot t, t' = t$$

and is known as the *Galilei transformation*. The relative velocities of the systems I and I' in this case are $v_{II'} = +v$ and $v_{I'I} = -v$. The generalisation of the *Galilei transformations* to three spatial dimensions does not provide any new problem and reads

$$x_k{}' = x_k - v_k t, \ t' = t.$$

The most general Galilei transformation which connects systems of coordinates in I and I' is composed of the pure Galilei transformation discussed here and the internal transformations in an inertial system. It depends on 10 independent parameters and reads

$$x_k{}' = a_{ki}x_i - v_k t + a_k, \quad t' = t + \beta \qquad \text{with} \qquad a_{ik}a_{lk} = \delta_{il} = a_{ki}a_{kl}$$

This Ends Our Digression on the Galilei Transformation

Within the context of special relativity, we have to determine not one but the three functions $k(v)$, $\mu(v)$, and $\nu(v)$ and that without the assumption of an absolute and universal time. This goal can be achieved if we make use of further properties of inertial systems. Since the constituting ensemble Γ does not distinguish any particular direction in space, we require again the postulates 4 and 5 and proceed in the following way:

Postulate 4: Isotropy of the transformation $T_{II'}$

If we consider – as always in the present context – only one spatial coordinate x, the requirement of *isotropy* means that a simultaneous change of signs of x, x', and v does not change the transformation. Hence, the original transformation

$$x' = k(v)(x - vt), \qquad t' = \mu(v)t + \nu(v) \cdot x$$

is changed by the transition

$$(x, x', v) \to (-x, -x', -v)$$

into the new transformation

$$x' = k(-v)(x - v \cdot t), \quad t' = \mu(-v)t - \nu(-v) \cdot x$$

which is not different from the original transformation. Comparing the coefficients of both transformations, we find

$$k(v) = k(-v), \quad \mu(v) = \mu(-v), \quad \nu(v) = -\nu(-v).$$

Instead of the *odd* function $\nu(v)$ we will use here the *even* function

$$\alpha(v) := -\nu(v)/v \cdot \mu(v).$$

Hence, we obtain for the transformation $T_{II'}$

$$x' = k(v)(x - v \cdot t), \quad t' = \mu(v)(t - \alpha(v), v, x)$$

with three even functions k, μ, α.

The constituting ensemble Γ does not distinguish one particular system of inertia. With respect to their defining properties, inertial systems are indistinguishable. We summarize this statement again in

Postulate 5: No system of inertia is distinguished with respect to the ensemble $\Gamma(k_1,k_2 \ldots k_n)$ (*principle of relativity*).

Remark: As already mentioned above, the principle of relativity is not an empirical principle but follows from the definition of an inertial system and the ensemble Γ. We emphasise again that this principle is not a characteristic of special relativity but it holds already in the Newtonian theory of space-time. Hence, also for the derivation of the Galilei transformations it had to be taken into account.

For the transformation between two inertial systems, the postulate 5 means that the transformations $T_{II'}$ and $T_{I'I}$ have the same formal structure. The application of this property allows for further specification of the transformation. Using again the denotation $v = v_{II'}$ and $v' = v_{I'I}$ for the relative velocities of I and I', we can compare the transformations $T_{II'}(v)$ and $T_{I'I}(v')$ and get

$$T_{II'}(v): \quad x' = k(v)(x - v \cdot t), \qquad t' = \mu(v)(t - \alpha(v)v \cdot x)$$

$$T_{I'I}(v'): \quad x = k'(v')(x' - v' \cdot t'), \qquad t = \mu'(v')(t' - \alpha'(v')v' \cdot x')$$

The indistinguishability, i.e. the equality of the formal structure implies

$$k(v) = k'(v), \quad \mu(v) = \mu'(v), \quad \nu(v) = \nu'(v).$$

If these conditions are fulfilled, then the functions $T_{I'I}\,(v')$ and $T^{-1}{}_{II'}(v)$ should generally agree. For evaluating this requirement we calculate from

$$T_{I'I}(v') : \quad x = k(v')(x' - v'.t'), \qquad t = \mu(v')(t' - \alpha(v')v'.x')$$

the relative velocity v

$$v = (dx/dt)_{dx'=0} = -k(v')v'/\mu(v') =: u(v')$$

and from

$$T_{II'}(v): \quad x' = k(v)(x - vt), \qquad t' = \mu(v)(t - \alpha(v)v \cdot x)$$

the relative velocity v'

$$v' = (dx'/dt')_{dx=0} = -k(v)v/\mu(v) =: u(v).$$

From these relations for v and $v\prime$ we obtain for the function u the functional equation

$$v = u(v') = u(u(v)).$$

If we require that u is defined in the entire velocity space, then the solutions $u(v) = \pm 1/v$ are ruled out. At first, the two solutions $u_1(v) = v$ and $u_2(v) = -v$ remain. However, if we require that for $v \to 0$ the transformation $T_{II'}(v)$ approaches continuously the identical transformation, then also $u_1(v)$ is ruled out and we obtain

$$v = v_{II'} = -v_{I'I} = -v'.$$

This result could also be obtained for Galilei transformations, whose derivation was finished at this point. Here, we have only

$$T_{II'}(v): \quad x' = k(v)(x - v \cdot t), \qquad t' = \mu(v)(t - \alpha(v)v \cdot x) \qquad \text{and}$$

$$T_{I'I}(v'): \quad x = k(-v)(x' + v \cdot t'), \quad t = \mu(-v)(t' + \alpha(-v)v \cdot x')$$

with three still undetermined even functions k, μ, α.

For further determining these functions, we make use of the equivalence of the transformations $T^{-1}{}_{II'}(v)$ and $T_{I'I}(-v)$. By comparing the coefficients of the time transformations

$$T^{-1}{}_{II'}(v): \qquad t = \frac{t'}{\mu(v)(1 - \alpha v^2)} + \frac{\alpha v x'}{k(v)(1 - \alpha v^2)}$$

$$T_{I'I}(-v): \qquad t = \mu(-v)t' + \mu(-v)\alpha v x'$$

we obtain

$$k(v) = \mu(v) = \frac{1}{\sqrt{1 - \alpha(v)v^2}}.$$

Hence we get for the transformation $T_{II'}(v)$:

$$x' = \frac{x - vt}{\sqrt{1 - \alpha(v)v^2}}, \qquad t' = \frac{t - \alpha(v)vx}{\sqrt{1 - \alpha(v)v^2}}.$$

For the determination of the function $\alpha(v)$ we apply again the principle of relativity, i.e. the indistinguishability of the systems I and I' and require

Postulate 6: The inertial systems I and I' are indistinguishable also with respect to a third inertial system I''.

Formally, this means that the iterative application of the transformations $T_{II'}(v)$ and $T_{I'I''}(v')$ with velocities $v = v_{II'}$ and $v' = v'_{I'I''}$ corresponds to the application of a third transformation $T_{II''}(v'')$ of the same type with $v'' = v_{II''}$. Using again as abbreviation $k(v) = 1/\sqrt{1 - \alpha(v)v^2}$ the connected transformation $T_{I'I''}(v') \otimes T_{II'}(v)$ reads

$$x'' = k(v')k(v)\{x(1 + \alpha(v)vv') - t(v' + v)\}$$

$$t'' = k(v')k(v)\{t(1 + \alpha(v')vv') - x(\alpha(v)v + \alpha(v')v')\}.$$

It should be possible to write these transformations in the form

$$x'' = k(v'')(x - v''t),$$

$$t'' = k(v'')(t - \alpha(v'')v''x).$$

If we compare the coefficients of this and the preceding form, we find

$$\alpha(v) = \alpha(v') = \alpha = \text{const.},$$

which means that α is a constant that does not depend on v.

Hence, for the velocity v'' we obtain

$$v'' = \frac{v + v'}{1 + \alpha vv'}$$

as the theorem for the addition of velocities.

The transformations can now be written in the form

$$T_{II'}(v, \alpha): \qquad x' = \frac{x - vt}{\sqrt{1 - \alpha v^2}}, \quad t' = \frac{t - \alpha vx}{\sqrt{1 - \alpha v^2}}$$

with a universal real constant α, whose sign and absolute value are still open.

For the determination of the sign of the constant α we make use of the requirement, that there are at least some pairs of events with the same temporal order in the systems I and I' with arbitrary relative velocity v. This requirement is not a new postulate, since we must require the invariance of the temporal order at least for those events, which correspond to points on the space-time trajectories of the test bodies of the ensemble Γ. Otherwise, the systems I and I' could be distinguished with respect to the trajectories of Γ, in contradiction to the above mentioned principle of relativity.

For evaluating this requirement, we distinguish two cases:

1. $\alpha =: -\eta^{-2} < 0$ (where η has the dimension of a velocity)
   In the transformation $T_{II'}(v, \alpha)$ all values of the relative velocity v are possible, i.e. all v-values with $-\infty < v < +\infty$ lead to real values of the coordinates (x', t').
2. $\alpha =: +\omega^{-2} > 0$ (where ω has the dimension of a velocity)
   In the transformation $T_{II'}(v, \alpha)$, the values of the coordinates (x', t') have real values only if the relative velocity v is restricted by the condition $-\omega < v < +\omega$ with $\omega > 0$.

These two cases are physically distinguished in the following sense.

1. The temporal distance $\Delta t = t_2 - t_1 > 0$ of two events $E_1(x_1, t_1)$ and $E_2(x_2, t_2)$ with the spatial distance $\Delta x = x_2 - x_1$ are transformed by a transformation $T_{II'}(v, \alpha)$ of type (I) according to

$$\Delta t' = \left(\Delta t + \frac{v\Delta x}{\eta^2}\right) \Big/ \sqrt{1 + \frac{v^2}{\eta^2}}$$

   Since v can assume any value, we can always find an inertial system $I'(v)$ such that $\Delta t' < 0$. In this case, the velocity v must merely fulfil the condition $|v| > \eta^2 |\Delta t|/|\Delta x|$. This means that for any pair (E_1, E_2) of events the chronological order can be changed by a convenient transformation to an other inertial system I'. Hence, there is no pair of events (E_1, E_2) with an (invariant) temporal order which is the same in all systems of inertia.
2. The temporal distance $\Delta t = t_2 - t_1$ of two events $E_1(x_1, t_1)$ and $E_2(x_2, t_2)$ with the spatial distance $\Delta x = x_2 - x_1$ are transformed by a transformation $T_{II'}(v, \alpha)$ of type (II) according to

$$\Delta t' = \left(\Delta t - \frac{v\Delta x}{\omega^2}\right) \Big/ \sqrt{1 - v^2/\omega^2}.$$

   Since the velocity v is restricted in this case by the condition $-\omega \leq v \leq +\omega$, the temporal order of two events $E_1(x_1, t_1)$ and $E_2(x_2, t_2)$ with $|\Delta x/\Delta t| < \omega$ cannot be changed. Hence, pairs of events $E_1(x_1, t_1)$ and $E_2(x_2, t_2)$ with $|\Delta x/\Delta t| < \omega$ have the same temporal order in all systems of inertia, i.e. their temporal order is invariant with respect to transformations of type (II). These transformations that depend except of v on the constant ω are denoted in the following text by $T_{II'}(v, \omega)$.

On the basis of these results, it is now easy to decide between the two possible signs of α. In case (I), $\alpha < 0$, there are no pairs (E_1, E_2) of events whose chronological order is the same in all systems of inertia. In case (II), $\alpha > 0$, there are at least some pairs (E_1, E_2) of events with an invariant chronological order. These pairs are characterised by $V := |\Delta x|/|\Delta t| \leq \omega$. According to the above formulated postulate, that at least for some pairs of events the chronological order should be the same in all inertial systems, we can decide now for $\alpha > 0$.

The corresponding transformation $T_{II'}(v, \omega)$, i.e.

$$x' = \frac{x - vt}{\sqrt{1 - v^2/\omega^2}}, \quad t' = \frac{t - \frac{vx}{\omega^2}}{\sqrt{1 - v^2/\omega^2}}$$

contains still one undetermined constant ω. On account of $-\omega < v < +\omega$, the constant ω is the maximal relative velocity between two inertial systems, i.e. we have always $v_{II'} = v \leq \omega$. For the special value $\omega = c$, we obtain the Lorentz transformation $T_{II'}(v, c)$, but without a justification for this special choice of ω. As long as the constant ω is not yet determined numerically, we call the transformation $T_{II'}(v, \omega)$ "*generalised Lorentz transformations*".

For the determination of the numerical value of the constant ω, we could use several different ways. Presumably, the most simple method is to introduce at first in the inertial system I, which is at rest, space-time coordinates that establish *Einstein-synchronisation* of spatially separated clocks – and to require that also the space-time coordinates of the moving inertial system $I'(v)$ are *Einstein-synchronised*. By means of this requirement and other methods[12] the undetermined constant ω in the generalised Lorentz-transformation $T_{II'}(v, \omega)$ can be determined in principle *empirically* with the result $\omega = c$, where c is the velocity of light in vacuum. In this way, we would arrive finally at the Lorentz-transformations $T_{II'}(v, c)$.

The *generalised Lorentz transformations* $T_{II'}(v, \omega)$ between two inertial systems I and I′ form a 10-parameter *Lie group*, the *Poincaré group*.[13] With this last step we also arrive at the space-time continuum of special relativity, the *Minkowskian* space-time M. The Minkowskian space-time is a 4 – dimensional manifold M that is equipped with a metric tensor $g_{\mu\nu}$. It can best be characterised by the line-element ds^2, which is invariant under generalised *Lorentz transformations* $T_{II'}(v, \omega)$. With Cartesian spatial coordinates x^i the line-element ds^2 reads

[12] Other empirical methods for the determination of ω are for instance the time dilatation of moving clocks or the increase of the inertial mass of moving bodies. More details are discussed in Section 2.4.

[13] Mittelstaedt (2006).

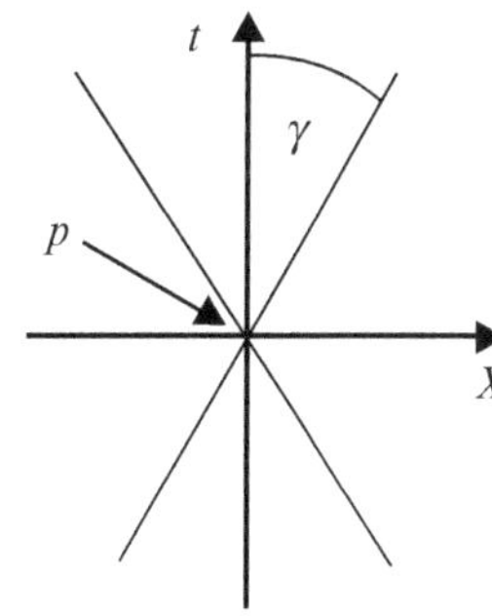

Fig. 2.1 The *null-cone* at point p and the cone angle γ

$$ds^2 = g_{\mu\nu}dx^\mu dx^\nu = \omega^2 dt^2 - (dx^1)^2 - (dx^2)^2 - (dx^3)^2$$

with $\mu, \nu, \in \{0, 1, 2, 3\}$ and $dx^0 = \omega\, dt$. The signature of $g_{\mu\nu}$ is 2 and thus $g_{\mu\nu}$ a *Lorentz metric.*[14] Within this indefinite Lorentz metric on a manifold M, the non-zero vectors at a point p can be divided into three classes, into vectors that are called *timelike, null,* or *spacelike*. In the space T_p of tangent vectors of M at p, the null vectors constitute a double cone, the *null cone*, which separates the timelike from the spacelike vectors. In the coordinates (x^1, x^2, x^3, t) the *null cone* can be expressed by the equation

$$(x^1)^2 + (x^2)^2 + (x^3)^2 - \omega^2 t^2 = 0$$

that is illustrated in (Fig. 2.1).

Hence, the apex angle γ of the *null cone* (from the t-axis) is connected with the constant ω by the relation $\gamma = \arctan \omega$.

The early attempts of a rational reconstruction of special relativity which we mentioned briefly in Section 2.1 can now be considered critically in the light of our present results. From a fundamental point of view, the attempt to reconstruct special relativity merely on the basis of the "principle of relativity" is irrelevant similarly as the attempt to base special relativity on two "axioms", the "principle of relativity" and the "principle of the constancy of the velocity of light". As to the latter principle, the so called second axiom, it became obvious that for a reconstruction of special relativity this postulate is not needed at all, except perhaps for the numerical determination of the constant ω, which could, however, also be obtained empirically. More important is the observation that also the first axiom, the "principle of relativity", cannot be used as an axiom since it follows from the definition

---

[14] Hawking and Ellis (1973), p. 38.

of the concept of an inertial system. For this reason, it holds equally in special relativity and in Newton's theory of space-time and must not be understood as a characteristic of Einstein's theory of special relativity.

2.3 Space-Time Intervals and Relativistic Mechanics

2.3.1 Measurements of Space-Time Intervals

According to the terminology of the last Section 2.2, we call points in the Minkowski space M again "events" $E(\boldsymbol{x}, t)$. We will consider here two events $E_1(\boldsymbol{x}_1, t_1)$ and $E_2(\boldsymbol{x}_2, t_2)$ with the spatial distance $\Delta \boldsymbol{x} = |\boldsymbol{x}_2 - \boldsymbol{x}_1|$ and the temporal distance $\Delta t = t_2 - t_1$.[15] Furthermore, for these two events E_1 and E_2 we can define the four-dimensional distance

$$(\Delta s)^2 = \omega^2 (\Delta t)^2 - (\Delta x)^2,$$

which can be positive, negative, or null – as the line element ds^2 defined in the last Section 2.2. Although the spatial distance $\Delta \boldsymbol{x}$ and the temporal distance Δt are changed by a generalised Lorentz transformation $T_{II'}(v, \omega)$, the four-dimensional distance $(\Delta s)^2$ remains invariant under these transformations. Indeed, if we consider for the sake of simplicity again only one spatial coordinate x, then from the transformation laws of a spatial distance

$$\Delta x' = (\Delta x - v\Delta t)\gamma(v) \qquad \text{with}\,\gamma(v) := 1/\sqrt{(1 - \mathrm{v}^2/\omega^2)}.$$

and of the temporal distance

$$\Delta t' = \gamma(v)\left(\Delta t - v\Delta x/\omega^2\right)$$

we obtain

$$\omega^2(\Delta t')^2 - \Delta\left(\boldsymbol{x}'\right)^2 = \omega^2(\Delta t)^2 - \Delta(\boldsymbol{x})^2$$

and thus

$$(\Delta s')^2 = (\Delta s)^2.$$

---

[15] Here and in the following we use bold letters for spatial three-vectors.

Hence, in particular, the property of $(\Delta s)^2$ being *positive*, *negative* or *null* is invariant.

Using this classification, pairs of events E_1 and E_2 can now be divided into three classes in an invariant way:

(I) If $|\Delta x(E_1, E_2)/\Delta t\,(E_1, E_2)| < \omega$ then the events E_1 and E_2 are *time-like*, and their chronological order is $T_{II'}(v, \omega)$ – invariant.

(II) If $|\Delta x(E_1, E_2)/\Delta t\,(E_1, E_2)| = \omega$ then the events E_1 and E_2 are *null*, and their chronological order is $T_{II'}(v, \omega)$ – invariant.

(III) If $|\Delta x(E_1, E_2)/\Delta t\,(E_1, E_2)| > \omega$ then the events E_1 and E_2 are *space-like*, and their chronological order is not $T_{II'}(v, \omega)$ – invariant.

Here we used the same terminology as for the three classes of *non-zero* vectors at a certain point p in the Minkowski space M.

On the basis of these results, we can now describe several "surprising phenomena" of special relativity. In the first years after the discovery of special relativity by Einstein, these surprising phenomena were the reason why special relativity was considered for almost 20 years as counterintuitive and not comprehensible. The results in question are, however, only very striking examples of phenomena that are merely not "directly intuitive" but merely after a detailed consideration "indirectly intuitive". We will try to show here that the mentioned phenomena are in fact "indirectly intuitive" and for this reason no longer surprising.

2.3.1.1 Relativity of Distant Simultaneity

Let us consider two events $E_1(x_1, t_1)$ and $E_2(x_2, t_2)$ at different places $x_1 \neq x_2$, that are, in a given inertial system I with coordinates $K_I\,(x, t)$, simultaneous, i.e. they have the same time values $t_1 = t_2$. According to the terminology just introduced, these two events are *space-like*. This means that a generalised Lorentz transformation $T_{II'}(v, \omega)$ to an other inertial system $I'(v)$, which is moving with velocity $v \neq 0$ and which has coordinates $K_{I'}(x', t')$, the chronological order of E_1 and E_2 can be changed such that for the transformed events $E'_1(x_1', t_1')$ and $E_2'(x_2', t_2')$ we get either $t'_1 < t'_2$ or $t'_1 > t'_2$. If the inertial system is realised by the earth, say, and the moving system I' by an aircraft, then the two events with the distance $\Delta x = x_1 - x_2$ and the vanishing time difference $\Delta t = t_1 - t_2 = 0$ in the earth system I, will have a non-vanishing temporal distance

$$\Delta t' = \gamma(v)\left(\Delta t - v\,\Delta x/\omega^2\right) \neq 0.$$

Obviously, for given values Δx and Δt in I, we could always find another system $I'(v)$ such that

$$\Delta t' > 0 \qquad \text{if } \Delta t > v\Delta x/\omega^2 \qquad \text{and}$$

$$\Delta t' < 0 \qquad \text{if } \Delta t < v\Delta x/\omega^{2\cdot}.$$

The result of these considerations is that for two distant events there is no absolute simultaneity. It is usually called "relativity of simultaneity".

This "relativity" can be illustrated by many surprising situations as it was shown in the famous book *Mr. Tompkins in Wonderland* by George Gamov.[16] However, the relativity of distant simultaneity is no longer surprising and counterintuitive if we realise that there is no absolute and universal time, which could be used for establishing an absolute simultaneity. Hence, in the light of our general program, the relativity of simultaneity is not *directly intuitive* but rather *indirectly comprehensible*, if we refer to the lack of absolute time.

2.3.1.2 Time Dilatation

If two events $E_1(x_1, t_1)$ and $E_2(x_2, t_2)$ are *timelike* to each other, i.e. if

$$|\Delta x/\Delta t| < \omega$$

then the chronological order of these events are invariant with respect to transformations $T_{II'}(v, \omega)$. Hence, $\Delta t = t_2 - t_1 > 0$ in the inertial system I implies that also $\Delta t' = t_2{}' - t_1{}' > 0$ in an other inertial system I'. We will assume here that this is the case. The time difference of E_1 and E_2 in the inertial system I' is then given by

$$\Delta t' = \gamma(v)\left(\Delta t - v\Delta x/\omega^2\right).$$

If the two events are in I at the same place $x_1 = x_2$ and thus $\Delta x = 0$, the relation between the two time intervals Δt and $\Delta t'$ reads

$$\Delta t' = \gamma(v)\Delta t.$$

By means of this relation, the phenomenon of time dilatation is expressed particularly clearly. Indeed, if Δt is the period of a clock C_I at rest in I at the place $x_1 = x_2$, and if Δt is thus a measure for the time interval between two events that correspond to the ticking of the clock C_I, then this period appears in the perspective of I' as dilated, i.e. we have

$$\Delta t' = \gamma(v)\Delta t \geqslant \Delta t.$$

[16] Gamov (1946).

The time intervals of a moving clock C_I (here the clock C_I at rest in *I*) appears to another observer (here at rest in I') as dilated, compared with the period of clock $C_{I'}$ of similar type at rest in I'. In other words, the moving clock C_I proceeds more slowly than the clock $C_{I'}$ at rest in I'. This phenomenon is usually called "*time dilatation*".

On account of the dynamical indistinguishability of inertial systems, i.e. on account of the principle of relativity, we can also invert the whole situation. A clock $C_{I'}$ at rest in I' with the period $\Delta t'$ in I', has in the coordinates of the inertial system *I*, which is moving with the velocity $(-v)$ relative to I', the period

$$\Delta t = \gamma(v)\Delta t' \geqslant \Delta t'.$$

In comparison to a clock C_I, of similar type, at rest in *I*, the moving clock $C_{I'}$ is again slower. Hence, we cannot state that one of the clocks C_I or $C_{I'}$ is slower than the other one, irrespective of the frame of reference in question, but only, that in the perspective of a given inertial system a moving clock is slower than a clock at rest of similar type.

2.3.1.3 Lorentz Contraction

The relativity of distant simultaneity has interesting and surprising implications for the measurement of spatial distances. For demonstrating this result we consider an inertial system *I* and a rigid measuring rod *S* at rest with the length $l_0 = x_2 - x_1$. The two end points of the rod l_1 and l_2 are then at the positions l_1: $x = x_1$, l_2: $x = x_2$.

If *S* is not at rest but moving with velocity $v \neq 0$, then we must determine its length by a convenient measurement process. Instead of moving the measuring rod *S*, we consider the measuring rod at rest from the perspective of another inertial system $I(-v)$ moving with the velocity $(-v)$. In this system $I(-v)$ the trajectories of the end points l_1 and l_2 read

$$I'(-v) \quad \begin{pmatrix} l_1 : x' = x'_1(0) + vt' \\ l_2 : x' = x'_2(0) + vt' \end{pmatrix}$$

The generalised Lorentz transformation between the inertial system *I* and I'

$$x = \gamma(v)(x' - vt'), \quad t'' = \gamma(v)(t' - vx'/\omega^2)$$

is determined such that at $t = 0$ there is coincidence of the two origins of the systems of coordinates.

If we consider the events

$$E_1 : x'_1 = x'_1(0),\ t_1' = 0; \qquad E_2 : x'_2 = x'_2(0),\ t_2' = 0$$

which indicate the beginning of the motion of the endpoints l_1 and l_2, then we obtain the x – coordinates of these events in the inertial system I by

$$x_1 = \gamma(v)x'_1(0), \quad x_2 = \gamma(v)x'_2(0)$$

which allow to determine the initial values $x'_1(0)$ an $x'_2(0)$. With this result we obtain the trajectories of l_1 and l_2 in its final form

$$I'(-v) \quad \begin{pmatrix} l_1 : x'_1(t') = x_1/\gamma(v) + vt' \\ l_2 : x'_2(t') = x_2/\gamma(v) + vt' \end{pmatrix}$$

The length l' of the measuring rod S in the inertial system $I'(-v)$ is defined as the difference of the x' – coordinates at the same time t, i.e.

$$l' := x'_2(t') - x'_1(t') = (x_2 - x_1)/\gamma(v) = l_0/\gamma(v)$$

where l_0 is again the length of S at rest in I. It is called rest length or proper length.

Hence, the length of the same S is contracted in any other moving system $I'(v)$ according to

$$l' = l_0/\gamma(v) \leq l_0$$

This phenomenon is called length contraction or Lorentz contraction of moving bodies.

The Lorentz contraction is again – similarly as the time dilatation – *a* relative phenomenon. From an observer in the inertial system I' the moving rod S (which is at rest in I) appears contracted, compared with a similar rod which is at rest in I'. On account of the (dynamic) indistinguishability of all inertial systems, the situation can also be inverted: The measuring rod S' which is at rest in I' with the proper length l_0, appears to an observer at rest in I (to whom the rod S' is moved with velocity $(-v)$) contracted in comparison to the rod S at rest in I.

2.3.2 Relativistic Kinematics and Dynamics

On the basis of the preceding derivation of the space-time structure of special relativity, in particular the generalised Lorentz transformations $T_{II'}(v, \omega)$, we can now begin to formulate the elements of the relativistic kinematics and dynamics. In a given inertial system I with coordinates K_I (t, x^i) with $i \in \{1, 2, 3\}$ we consider the space-time trajectory T_B of a body B, which is described by the function $x^k = f^k(t)$ with $k \in \{1, 2, 3\}$. The velocity of B is then given by

$$v^k = dx^k/dt = df^k(t)/dt.$$

By the relation

$$d\tau = (1/\omega)\left\{\omega^2 dt^2 - \left(dx^1\right)^2 - \left(dx^2\right)^2 - \left(dx^3\right)^2\right\}^{1/2}$$

$$= dt\{1 - v^2/\omega^2\}^{1/2} = ds/\omega$$

we define a new time element, the so called "proper time" $d\tau$, which is connected by the relation $ds = \omega d\tau$ with the invariant line element ds. Empirically, $d\tau$ is the time element of a clock C_B, that is comoving with the body B. Since $d\tau$ is the time element of a concrete clock, it has the same value in all systems of inertia. In other words, $d\tau$ is invariant with respect to generalised Lorentz transformations. This argument confirms from an operational view the formal result that $d\tau$ is invariant since it is connected with the line element by $ds = \omega\, d\tau$.

2.3.2.1 Relativistic Velocity

The behaviour of the velocity v^k under generalised Lorentz transformations $T_{II'}\,(v, \omega)$ is complicated. For this reason, it is more convenient to formulate the kinematical and dynamical concepts of relativistic mechanics not with respect to the coordinate time t, but with respect to the invariant time element $d\tau$ of the proper time, which is connected with the invariant line element by $ds = \omega\, d\tau$. The world-line (trajectory) of a mass point can then be expressed by a function

$$x^\mu = x^\mu(\tau), \quad \mu \in \{0, 1, 2, 3\}.$$

The four dimensional vector u^μ of the velocity will be defined here by

$$u^\mu := dx^\mu/d\tau$$

and thus

$$\begin{aligned} u^\mu &= \left\{\omega\, dt/d\tau, dx^i/d\tau\right\} \\ &= dt/d\tau\left\{\omega,(d\tau/dt)\left(dx^i/d\tau\right)\right\}, \qquad \mathrm{i} \in \{1, 2, 3\}. \end{aligned}$$

With $u^0 = dx^0/d\tau = \omega\, dt/d\tau = \omega \cdot \gamma(v)$ and $\gamma(v) := 1/\sqrt{(1 - v^2/\omega^2)}$ we obtain

$$u^2 = \left(u^0\right)^2 - \sum \left(u^i\right)^2 = \omega^2 \cdot \gamma(v)^2 - \gamma(v)^2 v^2 = \omega^2,$$

which means that $u^2 = \omega^2$ is an invariant quantity.

2.3.2.2 Relativistic Acceleration

In analogy to the definition of the relativistic velocity u^μ, we can define a concept of relativistic acceleration a^μ that refers to the proper time τ by

$$a^0 := du^0/d\tau = \gamma^4 v/\omega(dv/dt)$$

$$a^i := \left(\gamma^2 + \gamma^4 v^2/\omega^2\right)(dv/dt)$$

where *dv*/*dt* is the acceleration in the sense of classical mechanics. On trajectories of free particles the acceleration disappears, i.e. we have the equation of motion

$$d^2x^\mu/d\tau^2 = du^\mu/d\tau = 0$$

of a free particle. For all other trajectories the equation of motion reads

$$d^2x^\mu/d\tau^2 = A^\mu\left(x^\lambda, u^\lambda, \tau\right) \neq 0.$$

The term on the right hand side of this differential equation is usually decomposed into two parts, the "force" F^μ which represents the external influence on the body, and the invariant rest mass m_o which is a measure of the inertia of the moving body. With the *ansatz*

$$A^\mu\left(x^\lambda, u^\lambda, \tau\right) = F^\mu\left(x^\lambda, u^\lambda, \tau\right)/m_0$$

we obtain the relativistic equation of motion

$$m_0\ d^2x^\mu/d\tau^2 = F^\mu\left(x^\lambda, u^\lambda, \tau\right)$$

which has some similarity with the equation of motion in Newton's classical mechanics. However, it should be emphasised that there are no formal reasons for decomposing the function A^μ into two parts as $A^\mu = F^\mu/m_0$. This decomposition can be justified only by its practical success for the description of mechanical processes.

For the interpretation of the relativistic equation of motion we make use of the invariant quantity

$$u^2 = \left(u^0\right)^2 - \sum \left(u^i\right)^2 = \omega^2$$

and obtain

$$du^2/d\tau = 2\left(u^0a^0 - \sum u^i \cdot a^i\right) = 0$$

and by multiplication with m_0

$$u^0F^0 - \sum u^iF^i = 0$$

Furthermore, if we define

$$m := \gamma(v)m_0 \text{ and } f^i := F^i/\gamma(v),$$

then we obtain the 0-equation in the non-relativistic formulation

$$d(m\omega^2)/dt = \boldsymbol{v}\cdot\boldsymbol{f} \qquad (0-\text{equation})$$

and the three spatial equations

$$d(m\,\boldsymbol{v})/dt = \boldsymbol{f} \qquad (1,2,3-\text{equations})$$

In the non-relativistic limit $v/\omega \ll 1$ we obtain for these two equations

$$d(m_o v^2/2)/dt = \boldsymbol{v}\boldsymbol{f}.$$

i.e. the non-relativistic energy conservation theorem, and

$$d(m_o\boldsymbol{v})/dt = f$$

i.e. the non-relativistic equation of motion.

On account of this obvious analogy we denote

$m := m_0\,\gamma(v)$ as relativistic mass and

$E := m\,\omega^2$ as relativistic energy.

The relation $E = m\omega^2$ is not only a meaningful denotation for the quantity $m\omega^2$. It expresses the equivalence between the energy of a mechanical system and its inertial mass. This can be demonstrated in the following way. Let

$$p^\mu = m_0 u^\mu$$

the four-vector of the momentum with the components

$$\mu = 0: \qquad p^0 = m_0u^0 = m_0\gamma(v)\omega = m\cdot\omega = E/\omega$$

$$\mu = 1,2,3: \qquad \mathrm{p}^k = m_0u^k = m_0\gamma(v)v^k = v^kE/\omega^2.$$

Obviously, the quantity E has some properties of the energy in the sense of the non-relativistic mechanics. Indeed, it holds

$$dE/dt = d(m\omega^2)/dt = m_0\ \omega^2 d\gamma(v)/dt = \boldsymbol{v} \cdot \boldsymbol{f}. = dE_{kin}/dt$$

since $\boldsymbol{v} \cdot \boldsymbol{f} = dE_{kin}/dt$ holds in classical mechanics. Hence, up to a constant, E is a measure for the kinetic energy and we put

$$E_{kin} = m\omega^2 + \text{const.} = \mathrm{m}_0\ \gamma(v)\omega^2 + \text{const.}$$

If we require that for $v = 0$ the kinetic energy vanishes, then we find for the constant the value $(-m_0\ \omega^2)$ and get

$$E_{kin} = m\omega^2 - \mathrm{m}_0\omega^2, \qquad E = E_{kin} + m_0\omega^2$$

By means of the four-vector p^μ we can form the invariant quantity

$$p^2 := \left(p^0\right)^2 - (|\boldsymbol{p}|)^2 = E^2/\omega^2 - (|\boldsymbol{p}|)^2 = {m_0}^2\omega^2$$

and obtain the useful formula

$$p^0 = \sqrt{p^2 + |\boldsymbol{p}|^2} = \sqrt{{m_0}^2\omega^2 + |\boldsymbol{p}|^2}.$$

The three-velocity $v^i{}_P$ depends on the momentum $\boldsymbol{p}$ and can be expressed by

$$v^i{}_P = p^i \Big/ \sqrt{{m_0}^2 + |\boldsymbol{p}|^2/\omega^2}.$$

However, in the particular case of a particle which has rest mass $m_0 = 0$, as the photon, the three-velocity $v^i{}_P$ does not depend any longer on $\boldsymbol{p}$ and is given by

$$v^i{}_P = p^i \Big/ \sqrt{{m_0}^2 + |\boldsymbol{p}|^2/\omega^2} = p^i \Big/ \sqrt{|\boldsymbol{p}|^2/\omega^2} = \omega\, p^i/|\boldsymbol{p}|.$$

Furthermore, from

$$p^2 = \left(p^0\right)^2 - (|\boldsymbol{p}|)^2 = {m_0}^2\omega^2 = 0$$

it follows that the energy-momentum four-vector of a photon is a future oriented null vector p with the components

$$p^\mu = (|\boldsymbol{p}|, \boldsymbol{p}).$$

Conversely, since photons are moving on the null cone, the trajectory of a photon is given by $ds = 0$. In this case, the three-momentum $p^i = \omega\ m_0\ dx^i/ds$ is meaningful only, if also $m_0 = 0$ and thus we obtain merely the proportionality $p^i \sim dx^i$, where the factor of proportionality is undetermined.[17] These latter results about zero-mass particles will become of particular interest for the discussion in Section 4.2.

2.4 The Numerical Value of the Constant ω: The First Answer to the Problem

In the preceding sections of the present Chapter 2 we reconstructed special relativity, i.e. the Lorentz transformations and the elements of a relativistic mechanics merely by abandoning the hypothesis $O(C)^1$ of the existence of an absolute and universal time. In this derivation it became obvious that this reconstruction leads only to the generalised Lorentz transformation $T_{II'}$ (v, ω) and to the generalised relativistic mechanics, in which the numerical value of the constant ω is left open. There is no way for a theoretical determination of this numerical value. However, the value of the constant ω can be obtained by experimental means and we mention here several ways for an experimental determination of the constant ω.

From a methodological point of view, there are at least two somewhat different ways for an empirical determination of ω. Firstly, we can make use of an important step in the reconstruction of the generalised Lorentz transformation. Secondly, we can use the somewhat surprising effects of relativistic mechanics which we discussed in Section 2.3. We will begin with the step mentioned in the reconstruction of the Lorentz transformation.

2.4.1 *The Invariance of the Einstein-Synchronisation*

In a given inertial system I it is convenient to introduce a system of coordinates $K_I(x^i, t)$ which fulfils the requirement of Einstein synchronisation, often indicated by the synchronisation parameter $\varepsilon = ½$. As already mentioned above, two clocks $C(x_A)$ and $C(x_B)$ which are at rest at different places x_A and x_B, respectively, can be synchronised by a convention. Einstein's method to synchronise clocks by means of light signals is well known: At a time $t_A{}^{(1)}$ a light signal is emitted in x_A, reflected in x_B and received again in x_A at the time $t_A{}^{(3)}$. For the time $t_B{}^{(2)}$ of the refection in x_B we make use of the condition

$$t_B{}^{(2)} = t_A{}^{(1)} + \varepsilon\left(t_A{}^{(3)} - t_A{}^{(1)}\right)$$

[17] Sexl and Urbantke (1992), p. 69.

where ε is an undetermined real parameter $0 \leq \varepsilon \leq 1$ that must be determined by a convention. *Einstein-synchronisation* corresponds to the value $\varepsilon = ½$. This convention implies that the velocity of light in the two possible directions has the same value. For methodological reasons we emphasise that this convention does not imply that the metric of space-time depends essentially on the existence of light, which – in this case – would have an important constituting influence on the structure of space-time. The same convention could also be achieved by other methods, e.g. by slow motion of clocks.[18]

On the basis of these definitions and conventions, the determination of the numerical value of the constant ω can now be carried through in the following way: We start with the inertial system I, which is assumed to be at rest, space-time coordinates that establish *Einstein-synchronisation* of spatially separated clocks – and require, that also the space-time coordinates of a moving inertial system $I'(v)$ that can be arrived at by a generalised Lorentz transformation, are *Einstein-synchronised*. By means of this requirement, the undetermined constant ω in the generalised Lorentz transformation $T_{II'}(v, \omega)$ can be determined (in principle) *empirically* with the result $\omega = c$, where c is the velocity of light in vacuum.

Our second way refers to several "surprising effects" of special relativity which we discussed in Section 2.3. We mention here three different effects that can be used for an empirical determination of the unknown constant ω.

2.4.2 Time Dilatation

In an inertial system I we compare the period Δt of a clock C_1 at rest in x_1 with the period $\Delta t'$ of a clock C_2 at the place x_2 moving with velocity v. If the two clocks are at the same place, $x_1 = x_2$, then the two periods can directly be compared. In this situation we obtain the result

$$\Delta t' = \gamma(v, \omega)\Delta t \qquad \text{with} \qquad \gamma(v, \omega) = 1\Big/\sqrt{1 - v^2/\omega^2}.$$

From this equation we get

$$(\Delta t'/\Delta t)^2 = 1/(1 - v^2/\omega^2)$$

and in a few steps finally

$$\omega = v\Big/\sqrt{1 - (\Delta t/\Delta t')^2}.$$

[18] Mittelstaedt (1976/89).

Since the values of v, Δt, and $\Delta t'$ can be measured, the numerical value of ω can be determined by this equation with the result $\omega = c$, where c is again the velocity of light in vacuum.

2.4.3 Lorentz Contraction

In an inertial system I with coordinates K_I $(x^k_, t)$ we consider a measuring rod S at rest with the length $l_0 = x_2 - x_1$, where x_1 and x_2 are the coordinates of the two endpoints l_1 and l_2 of the rod. Furthermore, we consider another inertial system I' $(-v)$ moving relative to I with the velocity $(-v)$. If we measure the length of the measuring rod S in the moving system $I'(-v)$, – and that according to the well defined rules for the measurements of spatial distances of moving bodies – then we obtain a length l' that is smaller than l_0. For an observer O' in I' the measured length l' of S is shorter than the length l_0 of an identically constructed rod at rest in $I'(-v)$.

This "length contraction" of moving bodies is expressed in the present problem by

$$l' = l_0/\gamma(v, \omega) \leq l_0.$$

More explicitly this relation reads

$$l' = l_0\sqrt{1 - v^2/\omega^2} \leq l_0.$$

From this result we obtain

$$(l'/l_0)^2 = 1 - v^2/\omega^2$$

and finally the desired result for ω

$$\omega = v\sqrt{1 - (l'/l_0)^2}.$$

Since v, l_0 and l' can directly be measured, the value of ω can be determined as $\omega = c$ by this equation.

2.4.4 The Inertial Mass of a Moving Body

We consider an inertial system I. The inertial mass of a body B which is at rest in I is called its rest mass m_0. If we compare this mass $m_0 = m(0)$ with the inertial mass of another body B' moving with the velocity v, then we observe an increase of the inertial mass $m(v)$ such that

$$m(v) = m_0\,\gamma(v,\omega).$$

More explicitly, this relation reads

$$(m_0/m(v))^2 = 1\Big/\gamma(v,\omega)^2 = 1 - v^2\big/\omega^2$$

and thus

$$v^2/\omega^2 = 1 - (m_0/m(v))^2$$

and
$$\omega = v\Big/\sqrt{1-(m_0/m(v))^2} = c$$

In the four measurement procedures for the numerical determination of the constant ω, we could provide a first answer to the problem, why the numerical value of ω agrees with the velocity of light in vacuum. First of all, this is an empirical result. We emphasise again that this numerical agreement does not necessarily mean also a conceptual agreement between ω and c. A second answer to the problem mentioned will at least be indicated in Section 4.2 of Chapter 4 where we investigate the meaning of the constant "c" in physics. However, presently it seems to be more important to explicate and to illustrate the problem of the numerical value of ω than to solve it. Presumably, it is one of the open problems in contemporary physics.

2.5 Could Special Relativity Have Been Discovered Already by Newton?

2.5.1 A Pseudo-Historical Digression

On the basis of the results of the preceding investigations, in the present section we will discuss the purely hypothetical question whether the theory of special relativity, which was discovered by Einstein in 1905, could have been discovered already at an earlier time. More precisely, we ask whether at a certain time and on the basis of the scientific knowledge that was available at this time the theory of special relativity could have been discovered – not only a few years – but more than 200 years before Einstein, when Newton had just finished his *Principia* in 1687.[19] This pseudo-historical turn of the general idea of reconstruction of a given theory

[19] Mittelstaedt (2003).

was first formulated by Oskar Becker and applied to problems of the history of mathematics.[20]

In Chapters 1 and 2 of the present considerations we explained that in addition to the well known historical models for the development of physical theories, the model of a permanent accumulation of knowledge and Kuhn's model of the progress by "revolutions", there exists a third way that we explained here in detail. Indeed, sometimes it could happen that the progress in physics does not consist of the accumulation of new results, but of the reduction and elimination of prejudices that are tacitly incorporated in the old theory. This third model applies in particular to the theories of modern physics in the twentieth century, which can be reconstructed by eliminating metaphysical hypotheses of Newton's classical physics.

For this way of explaining the development of physical theories, classical physics and in particular classical mechanics plays an important role. Our ordinary experience (*OE*) which we briefly discussed in Section 1.1 is extended in many respects in classical mechanics and formulated in terms of mathematics. General principles and metaphysically motivated hypotheses were incorporated into this extended theory of ordinary experience in order to achieve a mathematically consistent and well established theory. The result of these efforts is Newton's theory of space-time and classical mechanics, which is presented in his *Principia*. It is loaded with several metaphysical hypotheses, the most important ones were formulated in Section 1.3.

It is a surprising observation that the reduction or elimination of these metaphysical hypotheses, without losing thereby the mathematical consistency of the theory, leads to the theories of modern physics. Our prime example for the reconstruction of a theory by systematic elimination of hidden metaphysical prejudices of classical mechanics is the reconstruction of the theory of special relativity, which we presented in the preceding Section 2.2. Among other interesting features, this reconstruction of (*SR*) shows how – in contrast to the actual historical development of this theory – the theory of special relativity could have been discovered, if step by step, scientists had freed themselves from the metaphysically motivated hypotheses of absolute time and Euclidean space which are contained in classical mechanics. Hence, our reconstruction of (*SR*) allows one presumably also to answer the question whether Newton could have discovered the theory of special relativity already at the end of the seventeenth century, say.

In the beginning of the twentieth century, the situation of special relativity was very confusing since it was not completely clear whether the theory provided new empirical or theoretical results in addition to several new conventions. In particular, the theory appeared to provide several surprising features, often considered as paradoxes, that contradict our common sense and our ordinary experience, which is usually considered as intuitive and comprehensible. For illustration, we mention

[20] Becker (1965), p. XIII f.

here 5 of the most important paradoxes of the theory, which are at first sight counterintuitive and not comprehensible. (For details we refer to Section 2.3.)

1. *There is no absolute simultaneity*
   It is not possible to say in an *absolute* sense that two events E_1 and E_2 happen at two distant points x_1 and x_2 in space, but at the same time values t_1 and t_2. If an observer $O(0)$ at rest finds simultaneity, i.e. $t_1 = t_2$, then another observer $O(v)$ moving with velocity $v > 0$ in another inertial system $I(v)$, will find out that the time difference between the two events reads $\Delta_v t = \gamma(v)\, v \cdot (x_2 - x_1)/\omega^2 \neq 0$.
2. *There is a time dilatation of moving clocks*
   In an inertial system I, we consider two observers O_1 and O_2 who are equipped with identically constructed clocks C_1 and C_2, respectively. If O_1 is at rest in I and O_2 moving with a constant velocity v, then in system I the moving clock C_2 runs slower than clock C_1. Hence, the period τ_2 of C_2 is larger than the period τ_1 of $C_1 : \tau_2 = \gamma(v) \cdot \tau_1 \geq \tau_1$.
3. *There is a clock paradox*
   We consider two observers O_1 and O_2 and assume that they are equipped with identically constructed clocks C_1 and C_2 which are perfectly shock-proof. The two observer meet together in ${x_A}^{(1)}$ at time ${t_A}^{(1)}$ (of clock C_1) and synchronize their clocks such that ${t_A}^{(1)} = {t_A}^{(2)}$ where ${t_A}^{(2)}$ is the time of clock C_2. Let O_1 be an inertial observer at rest and O_2 an observer who first moves away from O_1, and then changes the direction and moves back to O_1 in ${x_A}^{(1)}$. The moving observer reads on his clock C_2 a length of time τ_2 for the entire round-trip, whereas the observer O_1 measures on his clock C_1 a duration τ_1 for the round trip that is always larger than τ_2, i.e. $\tau_1 > \tau_2$. Hence, for the moving observer O_2 less time elapsed between the two encounters than for the observer O_1 at rest.
4. *There is a length contraction of moving bodies*
   Consider two inertial observer O_1 and O_2, where O_1 is at rest in an inertial system I_1 and O_2 is moving with constant velocity v. Both observers are equipped with identically constructed measuring rods for the measurement of spatial distances. If O_1 measures the spatial distance ${\Delta_{AB}}^{(1)}$ between two points x_A and x_B that are at rest in the system I_1, then the moving observer O_2 obtains for the distance of the same points the smaller value

   $$\Delta^{(2)}{}_{AB} = \Delta^{(1)}{}_{AB} \Big/ \gamma(v) \leqslant \Delta^{(1)}{}_{AB}.$$

   This reduction of distances is also called "length contraction".
5. *There is a maximal velocity of moving observers*
   Consider two inertial observers O_1 and O_2, where O_1 is at rest and O_2 is moving with constant velocity v – measured by O_1. This velocity v of O_2 can never exceed the limiting velocity ω, i.e. $v < \omega$, where ω is a universal constant that agrees numerically with the velocity c of light in vacuum. This means that the observer O_2 cannot move faster than the velocity of light.

These five statements are in contradiction to the corresponding statements of classical mechanics. In this context, classical mechanics plays a particular role, since it is not the theory of a certain domain of phenomena, but rather an "empty" theory, i.e. a theory, that deals with nothing but classical physics itself. In classical mechanics, we formulate the most general laws and structures which refer to any theory of classical physical objects. For this reason, the differences and contradictions of this theory with respect to special relativity, which we have just mentioned here, are of quite general importance.

The five paradoxical features of special relativity may illustrate among other things the reaction of the scientific community to Einstein's new theory of space and time. Hence, before we are going to answer the pseudo-historical question in the title of this section, we should clarify the significance of the purported paradoxes. At this point, we refer to the distinction between "directly intuitive" and "indirectly intuitive" which we discussed in Section 1.1. It is obvious, that the five facts in question are not directly intuitive but rather counterintuitive. They could, however, be "indirectly intuitive", i.e. reducible to "directly intuitive" facts via a sequence of logical steps. For our present problem, this is actually the case. Within the framework of Newton's theory of space-time, the paradoxical properties would completely disappear. However, we know that this classical space-time is loaded with several unjustified metaphysical hypotheses, in particular with the hypothesis of the existence of an absolute and universal time. Without this hypothesis, the resulting theory of space-time shows the paradoxical features mentioned. However, since the hypothesis of an absolute time is by no means intuitive and comprehensible, the consequences of an abandonment of the absolute time hypothesis are more intuitive and more comprehensible, i.e. the statements 1–5 are "indirectly intuitive" in the discussed sense. In other words, the five paradoxical statements are paradoxes in the literally sense and we expect that they can be resolved by a deeper inspection of the problem. Of course, this result must also be derived by the explicit reconstruction of special relativity.

The explicit reconstruction of special relativity, which we presented in Chapter 2, confirms this conjecture. In order to demonstrate that classical mechanics is based on hypothetical assumptions that clearly exceed our ordinary experience (*OE*), classical mechanics must be reformulated by the same conceptual and mathematical means as special relativity. This has been done in Section 2.2. On the background of this reformulated classical mechanics the theory of special relativity can be reconstructed merely by abandoning the ontological hypotheses in question. This reconstruction consists of several clearly distinguished steps, which will only briefly be mentioned here. All details can be found in Section 2.2.

The first step is the establishment of an Euclidean geometry in space which can be justified by the assumption that finitely extended measuring rods are freely movable. For the second step, the introduction of a convenient concept of time, we introduce the constituting ensemble $\Gamma(k_1, k_2, \ldots)$ of bodies k_i, that are freely thrust into space as well as a frame of reference equipped with measuring rods. If the trajectories of the test bodies k_i are Euclidean straight lines from the perspective of this frame of reference, then the frame of reference is called an inertial system I. While *topological* time can be tied to an arbitrary process, *metric* time is subject to the requirement that test bodies k_i not only move in a straight line but also at constant velocity. This requirement can be met since empirically the bodies k_i move uniformly relative to one another, which can be determined without knowledge of a metric time.

The third step deals with the combination of space and time in a space-time. It occurs through the definition of the synchronicity of spatially separated clocks and through the transformations between inertial systems. The internal transformations $I \to I$ of an inertial system I consist of the transformations of the Euclidean group and a *one* – parameter transformation for the time translation. On the basis of the above instructions, the transformations $I \to I'(v)$ from I to another inertial system $I'(v)$ moving at velocity v have (in *one* space coordinate) the form

$$x' = k(v)(x - vt), \qquad t' = \mu(v)t + \nu(v)x$$

with three arbitrary functions $k(v)$, $\mu(v)$, and $\nu(v)$.

At this point, the decision is made between classical mechanics and the theory of special relativity. If one presupposes the hypothesis $O(C)^1$ of the existence of an absolute time, then one would arrive in a few steps at the Galilei transformations

$$x' = x - v \cdot t \qquad t' = t$$

of classical mechanics. If the hypothesis $O(C)^1$ is completely abandoned, on the other hand, in a fourth step one obtains due to the kinematical indistinguishability of inertial systems the unknown functions $k(v)$, $\mu(v)$, and $\nu(v)$ and finally the generalised Lorentz transformation of special relativity $T_{II'}(v, \omega)$, i.e.

$$x' = \frac{x - vt}{\sqrt{1 - v^2/\omega^2}}, \qquad t' = \frac{t - \frac{vx}{\omega^2}}{\sqrt{1 - v^2/\omega^2}}$$

which still contains the undetermined constant ω. On account of the relation $-\omega < v < +\omega$, ω is the maximal relative velocity between two inertial systems, i.e. we have $v_{II'} = v \leq \omega$.

Since the constant ω appears in many physical processes, the numerical value of ω can be determined experimentally in various ways. Several methods were already mentioned at the end of the preceding section: In an inertial system I, which is at rest, we could introduce space-time coordinates that establish *Einstein-synchronisation* and require that also the space-time coordinates of a moving inertial system $I(v)$ that is obtained by a transformation $T_{II'}(v, \omega)$ are *Einstein-synchronised*. The result of this requirement is $\omega = c$. We could also use the time dilatation of a moving clock for the numerical determination of ω, or we could use the increase of the inertial mass of moving bodies. Hence, the *empirical* determination of the numerical value of ω does not provide any new problem.

On the basis of the preceding discussion we are now in the position to give a – perhaps preliminary – answer to the question in the title of this section, whether already Newton could have found the theory of special relativity. According to the general task formulated by *Oskar Becker* in 1965, which we already mentioned above, we must first ask for the scientific knowledge that was available at the time of Newton at the end of the seventeenth century and could have been used by him. From an experimental point of view, we mention here in particular the measurement of the velocity of light by O. Rømer in 1670, i.e. 16 years before Newton's *Principia* appeared in 1st edition. This measurement confirmed first of all that the velocity of light in vacuum is not infinite but has a finite limit which he determined explicitly. From a theoretical point of view we have seen that for the reconstruction of special relativity only the mathematical techniques for working with linear equations or collineations are required, i.e. techniques which were well known to Newton and his contemporaries.

By means of this empirical and theoretical knowledge all steps of our reconstruction in Section 2.2 could have been performed. Even without a universal time, the construction of inertial systems and the definition of a topological time was possible. The metric time could have been introduced by the requirement that the equation of motion should look as simple as possible. (In the real history of physics, this convention was found by *H. Poincaré* in 1898). Even the synchronisation of distant clocks by means of light signals in the sense of Einstein was already possible, since these signals were known to proceed with a finite velocity of light. Hence, all necessary means for a successful reconstruction of special relativity were already available to Isaac Newton.

In spite of these obviously good preconditions, the actual Isaac Newton could not have found special relativity. The reason is not that he did not possess the necessary mathematical and physical abilities, since from a formal point of view special relativity is a rather simple theory. But, presumably, he was too much biased by the ideas of the theology and metaphysics of his time and could, for that reason, not identify his own assumptions as unjustified ontological hypotheses. However, a critical reflection of this kind would have been the indispensable precondition of an abandonment or a relaxation of the hypothetical assumptions in question.

2.6 The Attempt to Reconstruct General Relativity

2.6.1 The Pseudo-Riemannian Character of Space-Time

The theory of general relativity was developed by Einstein after the formulation of special relativity during the period from 1907 through 1915. Einstein's journey from special to general relativity was far from being systematic and is characterised by various attempts in vain, by the application of several principles, hypotheses and changing goals. We will not go into the details of the history of this fascinating development of a new theory but instead refer to the literature, in particular to the contributions to this topic by J. Stachel.[21] For the development of general relativity Einstein made use of several physical and methodological principles but only of very few experimental results. As to the principles, we mention here the *equivalence principle*, the *requirement of general covariance*, the *principle of general relativity*, and at least in the first years of the general relativity project also *Mach's principle*. The result of these efforts is a well elaborated theory, which is meanwhile also confirmed by numerous experimental results. However, due to its rather confusing historical development it is very hard to say whether this theory is merely true with respect to experiments and observations, whether it is partly a priori true, or whether it is true in the sense of mathematics. In this situation, it is reasonable to perform a "rational reconstruction" of the theory in a similar way as we have already done it for the much simpler theory of special relativity in Section 2.2.

General relativity, as we understand it today, is a theory of space-time in the presence of gravitational fields and at the same time a theory of gravitation. As to the first part, the theory of space-time we note that for the development of special relativity (in Section 2.2) gravitational fields had to be excluded explicitly, since in the presence of gravitational fields finitely extended systems of inertia cannot properly be defined. Indeed, if we were starting again with a constituting ensemble Γ of bodies which are freely thrust into space, we would not find a frame of reference such that the trajectories of the test-bodies are straight lines in the sense of the Euclidean geometry. This remark does not refer to the theory of general relativity as such but merely to the well-known fact that gravitational fields cannot be screened off, which means that in the presence of gravitational fields there is no field-free region in space.

General relativity generalises the Minkowskian space-time of special relativity in several respects. General relativity leads to a Riemannian metric of the three-dimensional position space R_3 and to a pseudo-Riemannian space of signature 2 for the 4-dimensional space-time R_4. In addition, in its second part, general relativity connects the metric tensor of space-time and the Riemannian curvature tensor with the energy-momentum tensor of matter by means of Einstein's field equations. We will come back to this important aspect at the end of this section.

At first, we discuss the alleged unintuitiveness of general relativity. It refers, primarily, to the non-Euclidean or Riemannian character of the geometry of the 3-spaces and only in a second step to the pseudo-Riemannian structure of space-time. Obviously, it is difficult to discuss in detail the non-Euclidean character of the geometry of the 3-dimensional position spaces, since the geometry or metric of space cannot be perceived as such. It is useful, therefore, to focus on observable

[21] Stachel (2002).

preconditions from which these metric structures are derived. In this regard, Helmholtz' two theorems[22] are very helpful

1. *If finitely extended measuring rods are freely mobile in space, then the geometry measured with these rods is elliptical, hyperbolic or Euclidean.*
2. *If the free mobility is guaranteed only for infinitesimally extended measuring rods, then the geometry measured with these rods is Riemannian.*

The premise of the first theorem, the free mobility of finite measuring roads, will be called here *hypothesis* $H(E)$, whereas the premise of the second theorem, the free mobility of infinitesimally extended measuring rods, will be called *hypothesis* $H(R)$. Ordinary experience leads to an intuition of space that is certainly not strictly Euclidean in character. Even as the form of intuition as discussed by Kant in the *Critique of Pure Reason*, space is three-dimensional, topologically and presumably also metrizable, but no more than that. The Euclidean character of space is an additional assumption, which is introduced by classical mechanics and which presupposes the hypothesis $H(E)$ regarding the free mobility of finitely extended measuring rods. Within the domain of objects of classical mechanics, this hypothesis can certainly not be verified with the requisite precision. It is an empirically unverifiable additional assumption.

We obtain the spatial metric of the theory of general relativity not by abandoning the empirically unjustified hypothesis $H(E)$, as in the case of hypothesis $O(C)^1$ of the existence of an absolute and universal time, but rather by weakening it to the assumption $H(R)$ that only infinitesimally extended measuring rods are freely mobile, which yields the Riemannian character of the spatial metric according to Helmholtz' second theorem. This weakening is certainly compatible with the spatial intuition of ordinary experience, but not with the spatial geometry of classical mechanics. Hence, here too, it is the spatial intuition of classical mechanics expressed by $H(E)$ that is unintuitive, not the spatial geometry of the general theory of relativity obtained through weakening $H(E)$ to $H(R)$.

As a consequence of the Riemannian character of the 3-dimensional position space, any additional constructions and definitions referring to time and the connection of space and time can only be carried out locally. A constitutive ensemble Γ of bodies freely thrust into space can only be used locally and momentarily for the constitution of inertial systems, which are therefore called *local systems of inertia*. With these local inertial systems, it is nevertheless possible to define – locally and momentarily – a topological and a metrical time. The local inertial frames of reference also allow for local execution of the constructions required for the formation of space-time, whereby a Minkowskian space-time M_4 is constructed locally and momentarily at every point of the space-time. These pseudo-Euclidean M_4 spaces are tangent spaces of the finite space-time R_4, which thus reveals itself as a pseudo-Riemannian space R_4. Note, that this somewhat complicated way of

[22] Laugwitz (1960), pp. 145–149.

reasoning is necessary, since for the space-time R_4 with an indefinite metric there is no generalised Helmholtz theorem, which could be used in this case.

The measurement of space-time intervals as usually described by means of measuring rods and clocks is not only very laborious, but also unsatisfactory for methodological reasons, since clocks and measuring rods would enter the semantics of the theory as primitive entities. It is possible to avoid both drawbacks, however, if light beams and particle paths are used for measurement. The trajectories of particles as well as those of light rays are geodesics of the Riemannian metric and are thus themselves objects of the theory. An explicit execution of this program can be found in Marzke and Wheeler[23] (light clocks) and in the axiomatic system of Ehlers, Pirani and Schild,[24] which has been expanded later by the radar geometry of Schröter and Schelb.[25] We shall not pursue this topic further, since it is of no consequence for our problem.

2.6.2 *Local and Global Phenomena*

According to the rational reconstruction of the 4-dimensional pseudo-Riemannian space-time with signature 2 in the preceding Section 2.6.1 it is obvious that we should not expect new and surprising effects in the local domain. Indeed, this pseudo-Riemannian space-time agrees locally with the Minkowskian space-time of special relativity. New and surprising effects in this space-time were already discussed in the preceding sections, in particular in Section 2.5.1. However, in finitely extended regions of space-time we should be prepared to discover – in addition to the effects of special relativity discussed in Section 2.5.1 – new phenomena that are not in accordance with the ontology $O(C)$ of classical physics, in particular with the hypothesis $O(C)^2$. This is, of course, even more the case if we consider global phenomena of the entire space-time, which are subject to the large-scale structure of space-time.

Global effects of this kind are well known in the field of relativistic cosmology. Apart from the well established formalism of line elements ds^2, the global structures can be visualized most adequately by means of "Penrose diagrams". This mathematical technique applies a convenient combination of coordinate-transformations and conformal transformations to the infinitely extended space-time and maps it to finite two-dimensional pictures. Since the formalism of Penrose diagrams is somewhat elaborate and cannot be presented here, we refer to the literature, in particular to the monograph by Hawking and Ellis, Chapter 5.[26] In order to provide a short but more intuitive idea of the strange and in some sense surprising phenomena in the

[23] Marzke and Wheeler (1964).

[24] Ehlers et al. (1972).

[25] Schröter and Schelb (1994).

[26] Hawking and Ellis (1973).

global domain of the universe, the entire space-time or for short the "world", we mention a few generally intelligible examples. The world could be

- a static, everlasting universe (Einstein model)
- an expanding universe (Friedmann model)
- a rotating universe (Gödel model)
- a universe that allows for closed time-like world lines and thus for time travel (Gödel model)
- a universe with an infinite age (Eddington-Lemaître model)

2.6.3 *Einstein's Field Equations*

The theory of general relativity is not only a theory of the pseudo-Riemannian space-time R_4 but also a theory of gravitation. The influence of a Riemannian space-time – as the guiding field – on the movement of test particles and light rays is locally indistinguishable from the influence of a gravitational field. This equivalence was one of the most important motivating principles of Einstein. Hence space-time is also influenced by the material sources of a gravitational field. General Relativity describes this influence by means of Einstein's field equations

$$G^{\mu\nu} := R^{\mu\nu} - \frac{1}{2} R\, g^{\mu\nu} = -\kappa T^{\mu\nu}$$

in which the Einstein-tensor $G^{\mu\nu}$ of space-time is proportional to the energy-momentum tensor $T^{\mu\nu}$ of matter ($R^{\mu\nu}$ is the Ricci-tensor, $R = R_{\mu}{}^{\mu}$ the curvature scalar, and $g^{\mu\nu}$ the metric tensor). The factor of proportionality is the gravitational constant κ. Since for the Einstein-tensor $G^{\mu\nu}$ the identity $G^{\mu\nu};\nu \equiv 0$ holds, Einstein's equations imply also the equation of motion of the field creating matter $T^{\mu\nu};\nu = 0$.

At this point, the following problem of our reconstruction of general Rrelativity becomes obvious. The Riemannian character of space-time is not exclusively induced by the gravitational field and its material sources. Of course, according to Einstein's field equations the energy momentum tensor $T^{\mu\nu}$ provides a gravitational field and thus a guiding field for particles and light rays. However, also for vanishing $T^{\mu\nu}$ there are, except from the Minkowskian space-time, many non-trivial solutions of Einstein's field equations. These "vacuum solutions" represent an additional contribution to the Riemannian structure of space-time. Formally, the vacuum solutions can be characterised by the reduced field equations that follow from Einstein's field equations for vanishing matter, i.e. for $T^{\mu\nu} = 0$. The solutions of these reduced field equations $R^{\mu\nu} = 0$ are also called "Einstein-spaces" and were extensively discussed in the literature.[27]

[27] Petrow (1964).

Hence, the pseudo-Riemannian structure of space-time is partly induced by solutions of the vacuum field equations and partly by solutions of Einstein's field equations with non-vanishing matter. Locally, these two contributions cannot be distinguished. Formally, the Riemannian curvature tensor can be decomposed in two additive terms that correspond to the curvature of the vacuum and to the curvature of a matter induced space-time.[28] To further elucidate this point, we decompose the Riemannian curvature tensor $R_{\alpha\beta\gamma\delta}$ according to

$$R_{\alpha\beta\gamma\delta} = C_{\alpha\beta\gamma\delta} - \frac{1}{n-2}\left(g_{\alpha\gamma}R_{\beta\delta} - g_{\alpha\delta}R_{\beta\gamma} + g_{\beta\delta}R_{\alpha\gamma}\right) - \frac{R}{(n-1)(n-2)} \times \left(g_{\alpha\delta}g_{\beta\gamma} - g_{\alpha\gamma}g_{\beta\delta}\right)$$

Since for vanishing matter ($T^{\mu\nu} = 0$) we have $R^{\mu\nu} = 0$ and $R = 0$, the *Weyl tensor* $C_{\alpha\beta\gamma\delta}$ describes the curvature of a matter free space-time. Note that $C_{\alpha\beta\delta}{}^{\alpha} = 0$.

Consequently, also our way of reconstructing general relativity refers only to the pseudo-Riemannian space-time structure as such and not to its possible separation in two components with different origins – and thus not explicitly to its connection with matter. This means, that our way of reconstructing general relativity by weakening the metaphysical hypotheses of classical mechanics, leads in the present case from the strong hypothesis $H(E)$ of Euclidean geometry of the 3-dimensional space to the relaxed hypothesis $H(R)$ of a Riemannian geometry of the 3-dimensional position spaces and finally, as explained above, to a pseudo-Riemannian geometry of space-time. Since there is no way to distinguish the two contributions mentioned to this space-time, there is no way to say anything about the coupling between space-time and matter, i.e. about Einstein's field equations and the coupling constant κ. In other words, our way to reconstruct general relativity by relaxing metaphysical hypotheses of classical mechanics comes to an end at this point. A justification of Einstein's field equations is not in sight along this way.

Einstein's field equations, which represent the second important part of general relativity, are as such loaded with several hypothetical assumptions that are neither intuitive nor justified by rational arguments. However, in contrast to the theological and metaphysical hypotheses of Newton's classical physics, which were eliminated here, Einstein's equations are based on several mathematical and methodological assumptions. Here, we refer to the above mentioned *principle of equivalence*, the *postulates of general covariance* and *general relativity* and also the more mathematical requirement that the field equations in question should be *the most simple quasi-linear second order differential equations in the pseudo-Riemannian space-time* R_4. In a similar context, Wheeler[29] mentions even six different hypothetical approaches to Einstein's equations.

[28] Hawking and Ellis (1973), p. 85.

[29] Wheeler (1973).

1. Einstein's original derivation, based on the principles of equivalence and correspondence.
2. Élie Cartan's derivation, resting on the fact that the boundary of a boundary is zero.
3. The most compact derivation one knows, based on the idea that density of mass-energy governs curvature.
4. The derivation of Hilbert and Palatini, founded upon the principle of least action.
5. The derivation of Hojman, Kuchar and Teitelboim, that introduces the group-theoretical concept of "group of deformations of a spacelike hypersurface in spacetime".
6. The schematic derivation of Andrei Sakharov, founded upon the concept of "the metric elasticity of space". (*Cf. footnote 36 of Chapter 4*)

There is, however still another way to general relativity, in particular to Einstein's field equations, that should be briefly mentioned. Here, we think of the flat space-time approach, which formulates the theory of gravitation in the Minkowskian space-time of special relativity. We could start for this approach with the equation

$$\Delta\Phi = 4\pi k\rho$$

for the scalar potential Φ in Newton's theory of gravitation, where ρ is the density of the field creating matter and k Newton's gravitational constant. We can replace Newton's field equation by a Lorentz-invariant equation if we substitute the density ρ by the energy-momentum tensor $T_{\mu\nu}$ and the scalar field Φ by a symmetric tensor field $\psi_{\mu\nu}$. Taking into account that the energy of the free tensor field should be positive, then the most simple generalisation of Newton's potential equation reads

$$\left(\frac{1}{c^2}\frac{\partial^2}{\partial t^2} - \Delta\right)\psi_{\mu\nu} = f\left(T_{\mu\nu} - \frac{1}{2}\eta_{\mu\nu}T\right)$$

with the subsidiary condition $\psi_{\mu\nu}{}^{,\nu} = \frac{1}{2}\,\psi^{\lambda}{}_{\lambda,\mu}$, where $T = T_{\mu\nu}\eta^{\mu\nu}$ and f is a coupling constant.

It is obvious that this Lorentz invariant equation is not yet Einstein's field equation. However, a detailed investigation[30] shows, that it is equivalent to the linear approximation of Einstein's equation according to the coupling constant κ. In the limit ($n \to \infty$) of a power sequence of approximations according to f we would arrive at Einstein's equations, where the coupling constants are related by $\kappa = f^2$.

Summarizing this brief report, we find that the various approaches to Einstein's field equations mentioned, are based on several mathematical or methodological assumptions. Consequently, on account of these hypothetical components Einstein's

[30] Mittelstaedt (1976), pp. 64–66.

field equations are not justified in the same sense as the pseudo-Riemannian character of space-time.

2.6.4 Conclusion

The considerations of the present section led to a result that is quite different from the corresponding arguments in the context of special relativity. In Section 2.2 we could show that abandoning the hypothesis $O(C)^1$ of an absolute time in classical mechanics results in special relativity, which is, for this reason, closer to the empirical truth than classical physics. In general relativity, we must distinguish two different parts. The first part, the pseudo-Riemannian structure of space-time, follows if we eliminate again the hypothesis $O(C)^1$ of an absolute time and if we relax in addition the assumption $O(C)^2$ of the Euclidean geometry of the three-dimensional position space. The relaxation of the latter assumption leads after a long line of reasoning to the result that the 4-dimensional space-time is a pseudo-Riemannian space R_4 with signature 2. This is an obvious gain of knowledge, since the space-time R_4 is closer to the truth than the Minkowskian space-time M_4 of special relativity. However, for the second part of general relativity, Einstein's field equations, there is no argument in sight that could justify an increase of knowledge by these equations. The situation is even worse. Einstein's field equation are loaded with new formal hypotheses that are not justified by rational reasoning or empirical evidence. For this reason, there is no guarantee that the consequences of the field equations, in particular the large variety of rather strange space-time models, correspond to possible situations of real world.

Chapter 3
Reconstruction of Quantum Mechanics

3.1 The Historical Development of Quantum Mechanics

The historical development of quantum mechanics offers a rather heterogeneous picture, a large variety of interpretations, goals, and philosophical classifications. Quantum mechanics was understood as a theory in the spirit of positivism, operationalism, and empiricism, – to mention here only a few of numerous interpretations. None of the main protagonists of quantum mechanics, Bohr, Heisenberg, Schrödinger, and Pauli agreed completely about the understanding of the new theory, nor do individual scientists permanently represent the same philosophical position. As an outstanding example, we mention here Heisenberg, who changed his assessment of quantum mechanics several times.

In the early history of quantum mechanics, the theory was based by its founders, in particular by Niels Bohr, on three principles, that played an important role in the first years of the theory: the so-called quantum postulate, the correspondence principle, and the principle of complementarity. According to Bohr, the quantum postulate "*attributes to any atomic process an essential discontinuity, completely foreign to classical theories and symbolised by Planck's quantum of action*".[1] The principle of correspondence expresses the methodological requirement to searching for analogies between quantum theory and classical physics, in Bohr's words, it "*expresses our endeavours to utilize all classical concepts by giving them a suitable quantum-theoretical reinterpretation*".[2] The notion of complementarity has a negative meaning because it expresses an essential restriction of quantum systems:

> "*The very nature of quantum theory forces us to regard the space-time coordination and the claim of causality, the union of which characterizes the classical theories, as complementary but exclusive features of the description, symbolizing the idealization of observation and definition, respectively*".[3]

---

[1] Bohr (1934), p. 53.

[2] Bohr (1934), p. 8.

[3] Bohr (1934), pp. 54–55.

P. Mittelstaedt, *Rational Reconstructions of Modern Physics*, Fundamental Theories of Physics 174, DOI 10.1007/978-94-007-5593-2_3,

It is obvious that these requirements, i.e. the quantum postulate, the principle of correspondence, and the complementarity statement cannot be considered as axioms of quantum mechanics in the formal logical sense. In addition, they are neither intuitive and comprehensible nor justified by experimental evidence. Instead, they are mere reactions of the first generation of quantum physicists to the completely new situation they were confronted with. Hence, for an adequate understanding of the theory, it seems best way to forget about these "principles" and to reconstruct quantum mechanics exclusively on the basis of well understood rational aspects. In the present chapter, we will provide a rational reconstruction of quantum mechanics of this kind.

3.2 The Reduction of Ontological Hypotheses

As already mentioned above (Section 1.2) our way to reconstruct quantum mechanics is similar to the reconstruction of special relativity in Section 2.2. Also in quantum mechanics we will use a convenient relaxation of the ontological premises of classical mechanics listed in Chapter 1.3. In special relativity we had to relax merely the metaphysical hypotheses concerning the structure of space and time. It is obvious that for the reconstruction of quantum mechanics other ontological assumptions of classical mechanics must be taken into account. However, this is not the only difference between the two approaches. More important is a methodological difference. For the reconstruction of quantum mechanics our starting point is not a certain material structure of quantum objects, but the language and logic which is used for the descriptions of quantum physical phenomena. Hence, our first step consists of a convenient reduction of the classical ontology $O(C)$ already mentioned above (in Section 1.3) and the first formulation of a quantum ontology $O(Q)$. In a second step, we will investigate the implications of these reductions for establishing a formal language of quantum physics and in particular a formal logic of quantum physics, which is usually called quantum logic.

Modern physics, in particular quantum physics, is characterised not only by a huge number of new experimental results and by a well-elaborated theory, but essentially by a radical change of the underlying ontology. The new ontology reflects various new insights of quantum theory and will be called here "quantum ontology". It is still a controversial question what this quantum ontology looks like in detail. We mention here in particular the ontology of substances (objects), of properties, and of unsharp (fuzzy) properties. It should be emphasised that the discussed changes of the traditional ontology are not modifications that are induced by incorporating new empirical results. Instead, the new quantum ontologies can be obtained from the traditional (classical) ontology merely by relaxing and weakening of various assumptions and hypotheses of the classical ontology.[4]

[4] Mittelstaedt (2005).

Here we will briefly mention the most important changes. The classical ontology $O(C)$ assumes that there are individual objects S_i and that these objects possess elementary properties P_λ in the following sense. An elementary property P_λ refers to an object such that either P_λ or the counter property $\bar{P}_\lambda$ pertains to the system. An object is thoroughgoingly determined, i.e. with respect to each property P the object S possesses either P or the counter property $\bar{P}$. From these strong requirements it follows that all objects can be individualised by elementary properties if impenetrability is assumed as an additional condition. For objects S_i of the external objective reality the causality law and the law of conservation of substance hold without any restriction.

There are important objections against this classical ontology $O(C)$. The strict postulates of $O(C)$ are neither both intuitive and plausible nor can they be confirmed and justified by experimental means. The rigorousness of the assumptions mentioned exceeds the mere qualitative everyday experience and it exceeds the possibilities of experimental tests. In particular, the principle of thoroughgoing determination mentioned above has never been tested with an accuracy which would allow calling the result a principle. Hence, classical ontology is neither intuitive nor is it justified by experimental evidence. – Moreover, what is more important, the ontology $O(C)$ is not in accordance with quantum physics. A quantum mechanical object does never possess all possible elementary properties P_λ either positive or negative; it is not carrier of all possible properties. Instead, only a subset of all properties pertain to the system and can simultaneously be determined. These "objective" properties pertain to the object like in classical ontology. From these restrictions it follows that in quantum mechanics no strict causality can be established and that object systems cannot be individualised and re-identified by means of their objective properties.

We will not use these empirical results for formulating the ontology of quantum phenomena. However, we learn from these considerations that the classical ontology contains too much structure and too strong requirements compared with quantum physics. This observation offers the possibility to formulate the ontology $O(Q)$ of quantum physics by merely relaxing some hypothetical requirements of classical ontology $O(C)$. We note that no new requirements must be added to the assumptions of the classical ontology $O(C)$. Quantum ontology can thus be formulated as a relaxed version of the ontology $O(C)$ in the following way.

$O(Q)^1$: If an elementary property P pertains to a system, then a test of P leads with certainty to the result P. In addition, any arbitrary property can be tested at a given object with the result that either P or the counter property $\bar{P}$ pertains to the object system.

$O(Q)^2$: Quantum objects are not thoroughgoingly determined. They possess only a few elementary properties, either positive or negative. Properties that pertain simultaneously to an object are called "objective" and "mutually commensurable".

$O(Q)^3$: Since a quantum object is not thoroughgoingly determined, for quantum objects there is no strict causality law.

$O(Q)^4$: The lack of thoroughgoing determination and strict causality implies that quantum objects cannot be individualised and re-identified at later times.

This relaxed ontology $O(Q)$ can now be used as a starting point for establishing a formal language and logic of quantum physics. The quantum ontology has far reaching consequences for the possibilities of proving or disproving elementary (material) propositions and thus for the pragmatics of a formal language of quantum physics. This quantum pragmatics turns out to be a relaxation of a corresponding pragmatics of the language of classical physics, i.e. compared with classical pragmatics the new quantum pragmatics provides less possibilities for justifying and refuting propositions that attribute properties to an object system.

3.3 The Formal Languages of Classical Physics and of Quantum Physics

3.3.1 The Formal Language of Classical Physics

On the basis of quantum ontology and in the sense of the operational approach mentioned we can establish a pragmatics, semantics, syntax. We will start with the calculus L_i of intuitionistic propositional logic, since this calculus can be justified by operational means only, i.e. by dialogs or by proof-trees.[5] On the basis of elementary propositions that possess certain well-defined pragmatic properties, the calculus L_i of formal intuitionistic logic can be established. Within the context of quantum logic two pragmatic properties of elementary (material) propositions are of particular interest. First, we mention the property (v) of value definiteness, which means that an elementary proposition is either true or false in the sense of the assumed semantics. In L_i value definiteness is not presupposed but replaced by the weaker requirement [v] that leaves this question open. It is well known that in L_i the assumption (v) for elementary propositions implies that all finitely connected propositions (by the logical connectives $\wedge$, $\vee$, $\neg$, and $\rightarrow$) are also value definite in the sense of (v). This leads to the calculus L_c of classical, Boolean logic.[6]

More important for the present problem is the property (a) of "unrestricted availability". An elementary proposition A is called "available", if after a successful proof of A this proof result is still "available" after the proof or disproof of another, arbitrary proposition B. The decisive step from the calculus L_i of intuitionistic logic to the calculus L_{Qi} of intuitionistic quantum logic is the relaxation of the strong availability requirement (a) for elementary propositions to the weaker requirement of "restricted availability" [a]. By "restricted availability" we understand the assumption that the result of a proof or disproof of A is available (in a dialog or any other proof procedure) after a subsequent proof or disproof of another proposition B only, if A and B are mutually commensurable,

[5] Mittelstaedt (1976) and Stachow (1976).

[6] Mittelstaedt (1976).

i.e. if the commensurability proposition $k(A, B)$ can be shown to be true. Two propositions A and B are called to be "commensurable" if after a proof (or disproof) of A, and a subsequent proof (or disproof) of B, a repeated proof attempt of A leads with certainty to the previous result. For elementary propositions A and B "proof" means a verification of the respective property by measurement, whereas for connected propositions "proof" means success in a dialog game together with material proofs of elementary propositions.

In order to illustrate the importance of the pragmatic preconditions (v), [v] and (a), [a] in intuitionistic logic L_i and in intuitionistic quantum logic L_{Qi}, respectively, we derive a few of the most important formulas in both logical systems. Since first of all, there are no reasons to doubt in the value definiteness of elementary propositions in L_i and L_{Qi}, we can replace the more complicated proof procedures by means of dialogs as used in the literature[7] by the simpler and more intuitive methods of proof-trees.

The ontological preconditions mentioned guarantee the objective decidedness and finite testability of *elementary* propositions $A^e(S)$, which attribute *elementary* properties P^e(A) to an object system S. Hence $A^e(S)$ will be called to be *true*, $\vdash A^e$(S), if and only if the system S possesses the property $P^e(A)$. According to the preconditions the truth of $A^e(S)$ can be shown by a finite proof procedure, e.g. by a measuring process. In any case, either A or the counter proposition $\bar{A}$ turns out to be true in this way. This is meant by the statement, that elementary propositions are value definite. The semantics that is established by this concept of truth will be called "realistic".

In the following discussion we will consider the formal object language $S(S)$ of propositions $A(S)$ that attribute properties $P(A)$ to an object system S. In the formal language S_C of *classical physics* and on the basis of the subset $S_C^{(e)}$ of value definite elementary propositions we can introduce the logical connectives by the possibilities to attack and to defend them, i.e. by the possibilities to prove or to refute the connective. As an example we consider the two-place-operation "sequential conjunction"

connective	denotation	attacks	defences
$A \sqcap B$	„*A* and then *B*"	1.*A*?, 2.*B*?	1.*A*!, 2.*A*!

where A? means the challenge to prove A, and A! the successful proof. This attack- and defence scheme can be illustrated most conveniently by a proof-tree which is chronologically ordered.[8,9] The first branching point corresponds to the test of A at t_1, the second one corresponds to the B-test at t_2 (Fig. 3.1).

The ordinary straight line corresponds to a successful branch, the two dashed lines to branches without success. The chronological order is fixed here, but the time difference $\delta t = t_2 - t_1 > 0$ may assume arbitrary positive values. There is one branch of success.

[7] Mittelstaedt (1978).

[8] Mittelstaedt (1978).

[9] Stachow (1980).

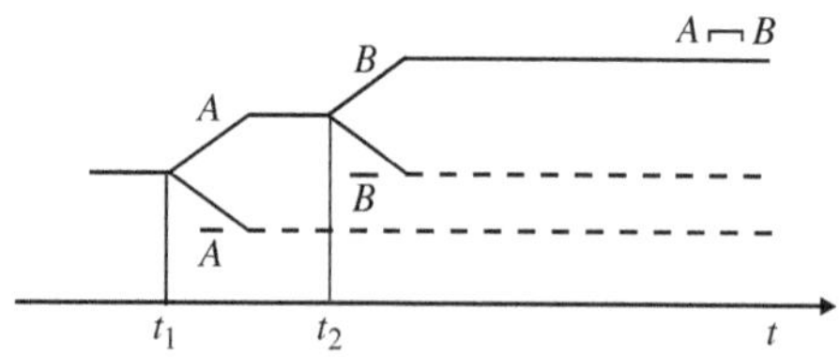

Fig. 3.1 Proof-tree for $A \sqcap B$ in classical language

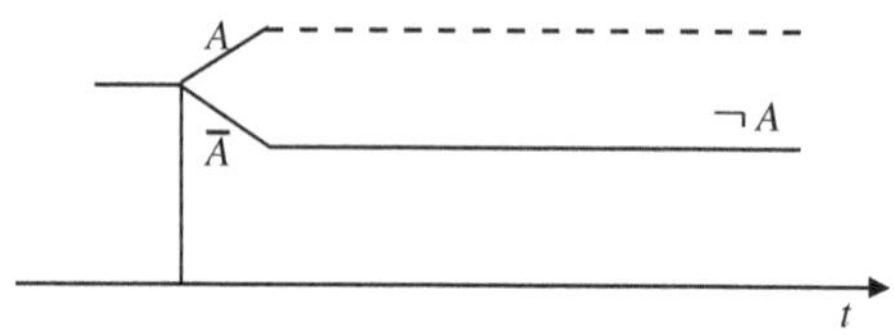

Fig. 3.2 Proof-tree for $\neg A$ in classical logic

In a similar way the one place operation "negation" may be introduced by the proof three for $\neg A$ (not A) with one branch for success and one branch without success (Fig. 3.2).

The sequential conjunction (and the other sequential connectives) refer to two instants of time t_1 and t_2. – The logical connectives refer to one common instant of time. The logical connective which corresponds to the sequential conjunction is the logical conjunction $A \wedge B$, which is defined by the following attack and – defence scheme

connective	denotation	attacks	defences
$A \wedge B$	A and B	$A?$, $B?$	$A!$, $B!$

In contrast to the sequential conjunction we have here an arbitrary number of attacks and defences in arbitrary order. In this way it is guaranteed that the result of the A and B – tests can be attributed to a common time value. If $A \wedge B$ is proved, then A and B are simultaneously true. The assumed independent testability of propositions A and B implies that after a test of B the result of a preceding test of A is still valid and available without any restrictions. The *unrestricted availability* of the results of A and B – tests implies that for the proof of $A \wedge B$ we need only two steps, provided their time difference $\delta t = t_2 - t_1$ is sufficiently small. There is no need in repeating the proofs (Fig. 3.3).

In a similar way the other *logical* connectives, i.e. the disjunction $A \vee B$ (A or B) and the material implication $A \rightarrow B$ (if A then B) can be defined by the possibilities to prove or to disprove them, or by the respective prove trees (Fig. 3.4).

The full language S_C of classical physics can then inductively be defined by the set $S_C^{(e)}$ of value definite elementary propositions and by all finitely connected

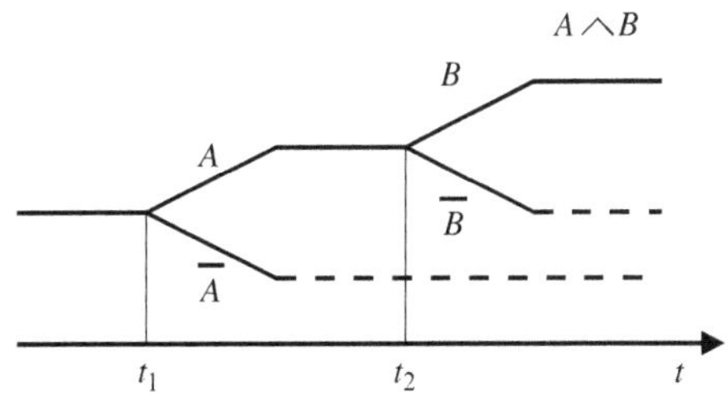

Fig. 3.3 Proof-tree for $A \wedge B$ in classical logic

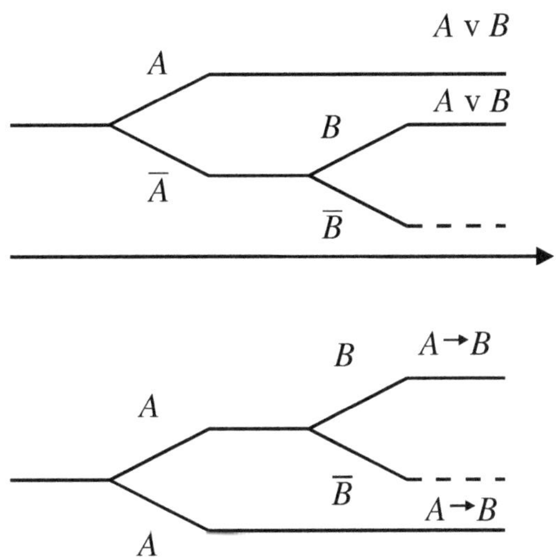

Fig. 3.4 Proof trees for $A \vee B$ and $A \rightarrow B$ in classical logic

compound propositions $A \in S_C$. The concept of truth can then be defined in the following way:

A proposition $A \in S_C$ is said to be true if the proof tree of A leads finally to a branch of success; it is called false if the proof tree ends with a branch without success.

Furthermore, on the set of propositions we introduce two binary relations,

1. the *value equivalence* $A = B$, A is true if and only if B is true
2. the *implication* $A \leq B \Leftrightarrow A = A \wedge B$

The full language can then be formulated as

$$S_C = \{S_C{}^{(e)}, \sqcap, \wedge, \vee, \rightarrow, \neg; =, \leq\}.$$

The semantics described here is a combination of a realistic semantics (with respect to elementary propositions) and a proof semantics (with respect to compound propositions). Hence, the truth of a compound proposition depends on the one hand on the connectives contained in it, on the other hand on the elementary propositions and their truth values. – This leads to the following question: Are there finitely connected propositions $A \in S_C$ which are true in the sense of the semantics described, irrespective of the truth values of the elementary propositions contained in them? Propositions of this kind will be called *formally true*. Examples for formally true propositions can easily be found. On account of the *value definiteness* of the elementary propositions we have the formally true proposition

$$A \vee \neg A \ \ (\textit{tertium non datur})$$

and on account of the *unrestricted availability* of propositions in a proof tree we have the formally true proposition

$$A \rightarrow (B \rightarrow A),$$

which is true even without the assumption of value definiteness of elementary propositions. If both *value definiteness* and *unrestricted availability* of propositions are assumed, then we obtain the formally true proposition

$$A \rightarrow ((A \wedge B) \vee (A \wedge \neg B)).$$

There are many, even infinitely many formally true propositions. The totality of formally true propositions is called classical logic and the algorithm which generates propositions of this kind is the *calculus* L_C *of classical logic*.

For the formulation of the calculus of the classical logic L_C we make use of two special propositions,

- the *true* proposition (*verum*) V, such that $\vdash V$ and for all $A \in S_C$ we have $A \leq V$ and
- the *false* proposition (*falsum*) $\Lambda = \neg V$ with $\Lambda \leq A$ for all $A \in S_C$. Using the two propositions V and Λ, we have $\vdash A \Leftrightarrow V \leq A$ and $\vdash A \rightarrow B \Leftrightarrow A \leq B$.

The calculus L_C can be formulated as a calculus of implications with "beginnings" $\Rightarrow A \leq B$ and rules like $A \leq B \Rightarrow C \leq D$. – The Lindenbaum-Tarski algebra of L_C is a complete, complemented and distributive lattice L_B (Boolean lattice). If it is freely generated by a finite number of elementary propositions it is also atomic and fulfils the covering law.

The calculus L_C of classical logic reads

1.1.	$\Rightarrow A \leq A$
1.2.	$A \leq B; B \leq C \Rightarrow A \leq C$
2.1.	$\Rightarrow A \wedge B \leq A$
2.2.	$\Rightarrow A \wedge B \leq B$
2.3.	$C \leq A; C \leq B \Rightarrow C \leq A \wedge B$
3.1.	$\Rightarrow A \leq A \vee B$
3.2.	$\Rightarrow B \leq A \vee B$
3.3.	$A \leq C; B \leq C \Rightarrow A \vee B \leq C$
4.1.	$\Rightarrow A \wedge (A \rightarrow B) \leq B$
4.2.	$A \wedge C \leq B \Rightarrow C \leq A \rightarrow B$
5.0.	$\Rightarrow \Lambda \leq A, \Rightarrow A \leq V$
5.1.	$\Rightarrow A \wedge \neg A \leq \Lambda$
5.2.	$\Rightarrow V \leq A \vee \neg A$

The formal propositional logic does not depend on the elementary propositions which are contained in the formally true propositions. However, the logic depends on the general preconditions under which proof processes are possible. In the present case the most important preconditions are the *finite decidability* of elementary propositions and the *unrestricted availability* in a proof-process. Formally true propositions are not true in an absolute sense. Their truth follows from the pragmatic preconditions of proving or disproving propositions. Only in this transcendental sense they are a priori true.

3.3.2 The Formal Language of Quantum Physics

In the preceding Section 3.3.1 about the formal language of classical physics, we mentioned already the weak pragmatic preconditions [v] and [a], which presuppose neither value definiteness (v) nor unrestricted availability (a), respectively, in their strict version. Since the strong preconditions (v) and (a) cannot be justified in general by rational reasoning or by experimental means, and in accordance with our most general method of relaxing or abandoning ontological hypotheses, we restrict our investigations to the relaxed and less hypothetical conditions [v] and [a]. On the basis of these weak pragmatic preconditions [v] and [a] a formal language S_Q of quantum physics and the calculus L_{Qi} of intuitionistic quantum logic can be established.

For the constitution of a quantum language we begin again with elementary propositions A which attribute a property $P(A)$ to an object system S. Accordingly, the proof of the elementary proposition A consists in a measurement of property $P(A)$ with positive outcome. The possibilities for quantum measurements allow for the assumption that after the measurement of $P(A)$ we obtain either a positive or negative result.[10] Hence, an elementary proposition A can either be proved

[10] Busch et al. (1996).

(result A) or disproved (result $\bar{A}$) and are thus value-definite. Furthermore, if after a successful proof of A, a new proof attempt for A is made, then one obtains again the result A, if the applied measurement is repeatable. However, if after a successful proof of A another elementary proposition B is proved, then a new proof attempt for proposition A will in general not lead to the previous positive result. Hence, two propositions A and B are in general not simultaneously decidable. This is only the case if the corresponding properties $P(A)$ and $P(B)$ are commensurable. In this case we will call also the propositions A and B "commensurable".

Elementary propositions A, B, ... are thus in general incommensurable, i.e. not simultaneously (jointly) decidable. If proposition A, say, was shown to be true, then after a proof attempt of B and irrespective of the result (B or $\bar{B}$), a new proof attempt of A will in general not lead to the previous result. Instead, this result is available after the B-test only if A and B are commensurable. In a sequence of proofs the results are only *restrictedly available,* where the restrictions are given by the violations of commensurabilities. For the definition of the connectives the restricted availability is very important.

These restrictions do not invalidate the definitions of the negation $\neg A$ and the sequential conjunction $A \sqcap B$ which are defined here by the same proof-trees as in the language S_C of classical language. The negation is defined by one proof attempt and the sequential conjunction by two subsequent proof attempts. In both cases the restricted availability does not matter since repeated proof attempts do not occur here. However, the restrictions do matter if one tries to define the other connectives.[11,12]

The logical conjunction $A \wedge B$ is defined here by the same attack and-defence scheme as in classical language. Since unrestricted availability is no longer given here, the proof-tree for

connective	denotation	attacks	defences
$A \wedge B$	A and B	$A?$, $B?$	$A!$, $B!$

$A \wedge B$ consists of an *infinite* number of steps and cannot be reduced to two steps as in classical language. If, however, A and B were commensurable, then it would again be possible to reduce the proof-tree to one A-proof and one B-proof. In order to achieve generally at a finite proof-tree we make use of the commensurability proposition $k(A, B)$ which is defined to be true if and only if A and B are commensurable. The counter proposition is denoted here by $\bar{k}(A, B)$.[13] The logical conjunction $A \wedge B$ is then true if in addition to A and B also $k(A, B)$ is shown to be true. Hence, we have a proof-tree with three subsequent tests at time values t_1, t_2, t_3.

[11] Mittelstaedt (1978).

[12] Stachow (1980).

[13] Mittelstaedt (1978).

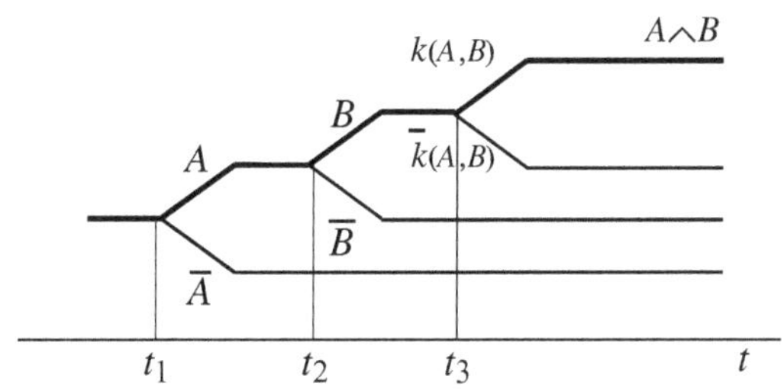

Fig. 3.5 Proof tree for the logical conjunction

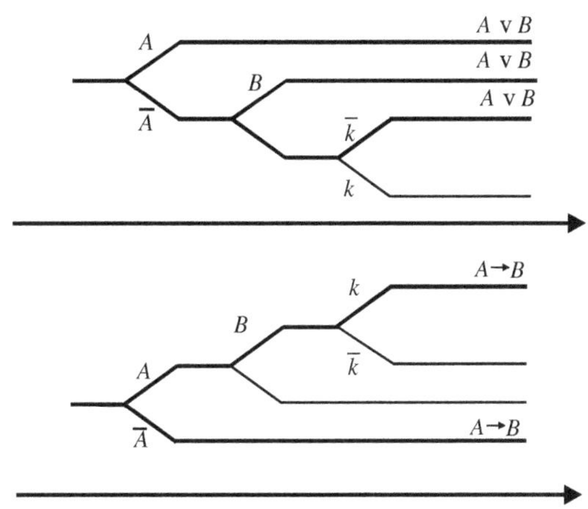

Fig. 3.6 Proof trees for the logical disjunction and the material implication

Since the conjunction $A \wedge B$ is understood as a simultaneous connective, the time differences t_3-t_2 and t_2-t_1 must be sufficiently small (Fig. 3.5).

The commensurability propositions $k(A, B)$ and $\bar{k}(A, B)$ are contingent propositions whose truth must be shown by a convenient sequence of measurements. We will not go into detail here. By means of the commensurability propositions $k(A, B)$ and $\bar{k}(A, B)$ one can define also the logical disjunction $A \vee B$ and the material implication $A \to B$ by proof trees with a finite number of steps. Similarly as in classical language we can define here binary relations between propositions. The proof equivalence $A \equiv B$ means that A can be replaced in any proof tree by B without thereby changing the result of the proof tree. The binary relation of value equivalence $A = B$ means that A is true if and only if B is true.[14] The relation of implication $A \leq B$ can then be defined by $A \equiv A \wedge B$. Finally, we mention that again $A \to B$ is true if and only if $A \leq B$ holds and that the commensurability proposition is true if and only if $A \leq (A \wedge B) \vee (A \wedge \neg B)$ holds (Fig. 3.6).

---

[14] If two propositions are *proof equivalent*, then they are also *value equivalent*. The inverse is not generally true. However, in classical language the two equivalence relations coincide.

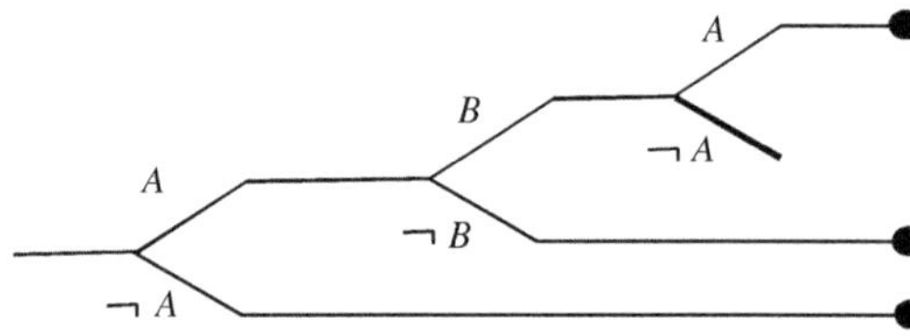

Fig. 3.7 Proof tree for $A \to (B \to A)$ in classical logic

The full language S_Q of quantum physics can then inductively be defined by the set ${S_Q}^{(e)}$ of elementary propositions, the commensurability propositions k and $\bar{k}$ and the connectives mentioned. Together with the relations "$\equiv$", "$=$", and "$\leq$" the language S_Q reads

$$S_Q = \{{S_Q}^{(e)}; k, \quad \bar{k}; \sqcap, \wedge, \vee, \to, \neg; \equiv, =, \leq\}$$

Quantum logic is the formal logic of quantum language S_Q and its syntax. The reduced possibilities of proving propositions are in particular important for those propositions which are true, irrespective of the elementary propositions contained in them, i.e. for formally true propositions. It turns out that in quantum language there are less formally true propositions than in classical language. In order to make this more preliminary information more precise we will express the totality of all formally true propositions of quantum language by a calculus, the calculus of quantum logic.

There are, first of all, many formally true propositions of classical language which are also formally true in quantum language. The value definiteness of elementary propositions implies that also all finitely connected propositions are vale definite, i.e. the proposition $A \vee \neg A$, the *tertium non datur law*, is formally true. The precondition that measurements are repeatable in principle implies that $k(A, A)$ is always true and hence $A \to A$, the *law of identity*, is formally true. In a similar way, it follows that $\neg(A \wedge \neg A)$, the *law of contradiction*, is formally true in quantum logic. The three cases mentioned are not very surprising since these formally true propositions contain only one proposition A. Hence, commensurability problems cannot appear. There are, however, also formally true propositions in quantum logic which contain two ore more elementary propositions, where nothing is presupposed about their mutual commensurability. An example of this kind is the proposition $(A \wedge (A \to B)) \to B$, the *modus ponens law*, which is formally true in quantum logic irrespective of the truth or falsity of the commensurability proposition $k(A, B)$.

More important for the characterisation of quantum logic are those propositions which are formally true in classical logic but not in quantum logic. The shortest and in addition most important proposition which is formally true in classical logic but not in quantum logic is the proposition $A \to (B \to A)$. In classical logic the proof-tree for $A \to (B \to A)$ contains only branches of success (Fig. 3.7).

In quantum logic the situation is more complicated since the proof tree for the material implication contains also the test of commensurability propositions $k(A, B)$. For this reason, the proof-tree for $A \to (B \to A)$ contains five branches, but

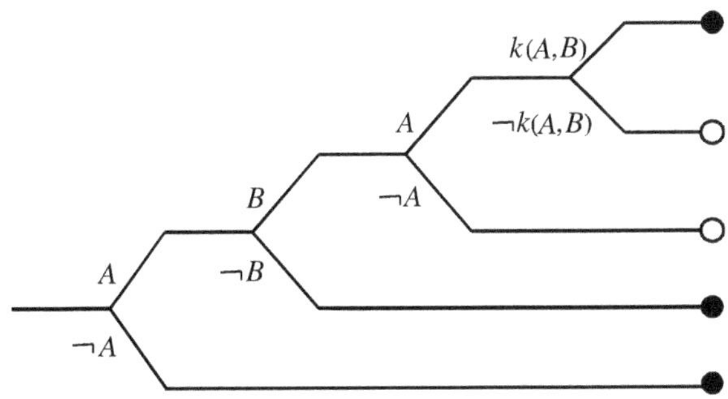

Fig. 3.8 Proof-tree for $A \rightarrow (B \rightarrow A)$ in quantum logic

only three branches of success. Only if the commensurability of A and B were presupposed, then the proof-tree would contain only successful branches. This means that in general the proposition $A \rightarrow (B \rightarrow A)$ is not true and thus not formally true (Fig. 3.8).

3.4 The Approach to Orthomodular Quantum Logic

On the basis of the weak pragmatic preconditions [v] and [a] of a formal language S_Q of *quantum physics*, which presuppose neither value definiteness nor unrestricted availability, we can establish now the calculus L_{Qi} of formal intuitionistic quantum logic. This calculus summarizes the totality of all propositions that are formally true even under the restrictions that are induced by the commensurability tests. There are – as in classical logic – infinitely many propositions that are formally true, i.e. true irrespective of the truth or falsity of the elementary propositions contained in them. They can be summarised in the quantum logical calculus L_{Qi}, which we present here as a calculus of implications $A \leq B$. It contains "beginnings" $\Rightarrow A \leq B$ and "rules" of the form $A \leq B \Rightarrow C \leq D$. For the formulation of this calculus we use of the two special propositions V (*verum*) and Λ (*falsum*) such that for all propositions $A \in S_Q$ the relations $\Lambda \leq A \leq V$ hold. – If $A \rightarrow (B \rightarrow A)$ is true then the relation $A \leq (B \rightarrow A)$ holds. $A \leq B \rightarrow A$ implies $B \leq A \rightarrow B$ and *vice versa* and $A \leq B \rightarrow A$ holds if and only if $k(A, B)$ is true. Hence, in a calculus of quantum logic the commensurability propositions $k(A, B)$ can be eliminated by this implication and will no longer appear in its final formulation.

The calculus L_{Qi} of intuitionistic quantum logic reads:

1.1.	$\Rightarrow A \leq A$
1.2.	$A \leq B; B \leq C \Rightarrow A \leq C$
2.1	$\Rightarrow A \wedge B \leq A$
2.2.	$\Rightarrow A \wedge B \leq B$
2.3.	$C \leq A; C \leq B \Rightarrow C \leq A \wedge B$

(continued)

3.1.	$\Rightarrow A \leq A \vee B$
3.2.	$\Rightarrow B \leq A \vee B$
3.3.	$A \leq C; B \leq C \Rightarrow A \vee B \leq C$
4.1.	$\Rightarrow A \wedge (A \rightarrow B) \leq B$
4.2.	$A \wedge C \leq B \Rightarrow A \rightarrow C \leq A \rightarrow B$
4.3.	$A \leq B \rightarrow A \Rightarrow B \leq A \rightarrow B$
4.4.	$B \leq A \rightarrow B; C \leq A \rightarrow C \Rightarrow B * C \leq A \rightarrow B * C; * \in \{\wedge, \vee, \rightarrow\}$
5.0.	$\Rightarrow \Lambda \leq A, \Rightarrow A \leq V$
5.1.	$\Rightarrow A \wedge \neg A \leq \Lambda$
5.2.	$A \wedge C \leq \Lambda \Rightarrow A \rightarrow C \leq \neg A$
5.3.	$A \leq B \rightarrow A \Rightarrow \neg A \leq B \rightarrow \neg A$

We mention briefly some important properties of this calculus of intuitionistic quantum logic:

(i) The calculus L_{Qi} is complete and consistent with respect to the dialog semantics mentioned,[15] i.e. all implications that are derivable in L_{Qi} can successfully be defended in a dialog and vice versa.

(ii) If the elementary propositions considered are value definite in the sense of (v), then all finitely connected propositions are also value definite. For these propositions the calculus L_{Qi} can be extended by the "excluded middle" $V \leq A \vee \neg A$. The extended calculus L_Q will be called the calculus of *orthomodular* quantum logic since the Lindenbaum-Tarski algebra of L_Q is an *orthomodular* lattice.[16] Compared with the calculus L_{Qi} the calculus L_Q contains only the new additional beginning

$$(5.4.) \Rightarrow V \leq A \vee \neg A.$$

The assumption that elementary (material) propositions are value definite is not purely hypothetical. It is based on the conjecture that the respective properties can be verified or falsified by material, experimental processes. It can easily be seen that L_Q is a relaxation of the calculus L_C of classical (Boolean) propositional logic.

(iii) The commensurability of two propositions A and B can be expressed in L_{Qi} by a binary relation $A \sim B$ given by $A \leq B \rightarrow A$ which is according to L_{Qi} symmetric. In L_Q this relation reads $A \leq (A \wedge B) \vee (A \wedge \neg B)$. The commensurability proposition mentioned above can be expressed in L_Q by

[15] Stachow (1976).

[16] Mittelstaedt (1978), p. 29, Kalmbach (1983).

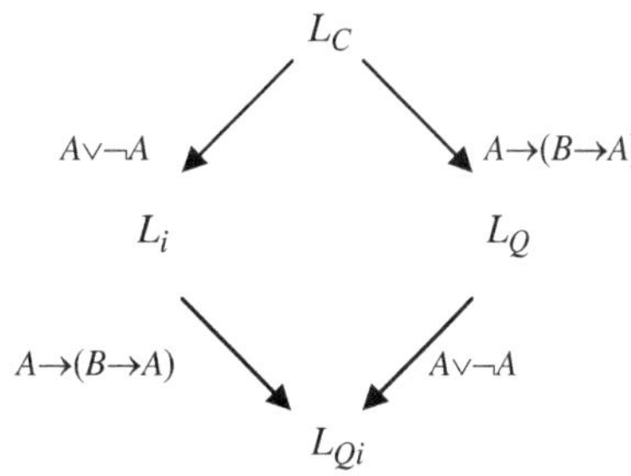

Fig. 3.9 Interrelations between the logical calculi L_{Qi}, L_Q, L_i, and L_C

$k(A, B) = (A \wedge B) \vee (A \wedge \neg B) \vee (\neg A \wedge B) \vee (\neg A \wedge \neg B)$. It is true if and only if $A \sim B$ holds. If $A \sim B$ is added to L_{Qi} as an additional general beginning,

$$(4.5.) \Rightarrow A \leq B \rightarrow A,$$

then one obtains the calculus L_i of intuitionistic logic. If it is added to L_Q, then one obtains the calculus L_C of classical logic.

(iv) If a set $\{A_1 \ldots A_n\}$ of elementary propositions is mutually commensurable, then the commensurability is inherited to all finitely connected propositions. Within the framework of the calculus L_Q of orthomodular logic this subset of mutually commensurable propositions is determined by a calculus L_C of classical logic.

(v) In the calculus L_{Qi} for two propositions A and B the material implication $A \rightarrow B$ is uniquely determined by L_{Qi}. $A \rightarrow B$ is true if and only if $A \leq B$ holds. In the extended calculus L_Q the uniquely defined material implication can be expressed by the other connectives as $A \rightarrow B = \neg A \vee (A \wedge B)$.

Summarising the various interrelations between the logical calculi L_{Qi}, L_Q, L_i, and L_C we obtain the following diagram (Fig. 3.9)

The strongest calculus L_C is on the top of the diagram, the weakest calculus L_{Qi} at the bottom. The arrows indicate the various relaxations leading from one calculus to another one. The propositions by the side of the diagram indicate the difference between the respective calculi. E.g. the proposition $A \rightarrow (B \rightarrow A)$ is formally true in L_i but not in L_{Qi}.

The explicit construction of the logical calculi L_{Qi} and L_Q is only one aspect of quantum logic. The Lindenbaum-Tarski algebras of L_{Qi} and L_Q are lattices L_{Qi} and L_Q, respectively and these lattices are of particular interest. Since the lattice L_Q is orthomodular we called L_Q the calculus of orthomodular quantum logic. If all propositions refer as predicates to a single object, then L_Q is *atomic* and fulfils the *covering law*.[17]

[17] Stachow (1984).

Definition: In a lattice L_Q an element α with $\alpha \neq \Lambda$ is called an *atom*, if for any element $X \in L_Q$, $\Lambda \leq X \leq \alpha$ implies either $X = \Lambda$ or $X = \alpha$.

Atomicity: If for any element $A \in L_Q$ there exists an *atom* α with $\alpha \leq A$, L_Q is called *atomic*.

Covering law: Let $\alpha \leq L_Q$ be an *atom*. If for all elements A and X of L_Q, the relation $A \leq X \leq A \vee \alpha$ implies $X = A$ or $X = A \vee \alpha$, the lattice L_Q is said to fulfil the *covering law*.

A lattice L_Q that is atomic and fulfils the covering law will be denoted here by L_Q^*.

3.5 The Bottom-Up Reconstruction of Quantum Mechanics in Hilbert Space

At this point, the two faces of quantum logic become obvious. Orthomodular quantum logic is not only a consequence of an ontology $O(Q)$ that is – at first sight – in accordance with quantum theory, it is presumably also the origin of this theory in the following sense. If we have once achieved an orthomodular lattice L_Q^* in the described way, it seems to be possible to proceed to the lattice L_{H} of subspaces of Hilbert space. The last step was strongly motivated by the Piron-McLaren theorem[18] which states that a lattice L_Q* (of length at least 4) is isomorphic to the lattice $L_H(D)$ of subspaces of a Hilbert space over a division ring D, where D is given by the real, the complex, or the quaternion numbers. If the real and the quaternion numbers could be excluded by experimental reasons, we would arrive at the Hilbert space $H(C)$ over the complex numbers C and thus at quantum mechanics in Hilbert space. If this way of reasoning were really conclusive, it would allow for testing the consistency of the reconstructed quantum mechanics with the underlying ontology $O(Q)$.

However, the lattice L_Q* does not restrict the choice of the division ring per se to the real, the complex, and the quaternion numbers. Indeed, Keller[19] could show in 1989 that there are lattices L_Q* that fulfil all the conditions of the Piron-McLaren theorem but nevertheless allow for non-classical Hilbert spaces over non-Archimedean division rings. This negative result was considered by many scientists as demonstrating the fundamental impossibility of the quantum logic approach to quantum mechanics in Hilbert space.[20] However, this discouraging conclusion was

[18] Cf. Piron (1964), MacLaren (1965), and Varadarajan (1968).

[19] Keller (1980).

[20] For more details cf. Dalla Chiara et al. (2001, pp.48–50, 2004, pp.72–74).

again disproved by an important result discovered by Maria Solèr[21] in 1995 that allows for a purely lattice-theoretical characterisation of classical Hilbert spaces. In fact, every lattice which satisfies in addition to the conditions of the Piron-McLaren theorem[22] also the so-called "angle bisecting condition" is isomorphic to a classical Hilbert lattice, in particular to the Hilbert space $H(C)$ over the complex numbers C. Obviously, we must confirm that the properties of quantum mechanics in Hilbert space are compatible with the ontological preconditions [a] and (v) of this approach.

Of course, this consistency requirement is fulfilled with respect to the commensurability problem, i.e. to the condition [a]. But it cannot be fulfilled with respect to the value definiteness, i.e. to the condition (v). Indeed, this ontological precondition of the preceding section is still too strong and cannot be realised in Hilbert space quantum physics. Orthomodular quantum logic is based on intuitionistic quantum logic L_{Qi} and the assumption that the material elementary propositions are value definite, since the respective properties can be verified or falsified by material processes. This assumption which allows to extend the calculus L_{Qi} to the stronger calculus L_Q is usually justified by a recourse to quantum mechanical measurements. However, more detailed investigations of the quantum theory of measurement have shown in recent years that after a unitary measurement process a definite value of the measured property cannot be attributed to the object system and no definite value can be attributed to the pointer of the measuring apparatus.[23,24] Consequently, elementary (material) propositions that attach truth-values to directly measurable properties cannot be value definite – in contrast to the assumption mentioned above.

One way to solve this "problem of objectification" consists in a relaxation of the underlying ontology $O(Q)$. Indeed, we could weaken the ontological presuppositions by considering unsharp properties and unsharp elementary propositions, which are not value definite. This idea is strongly supported by physics, since the quantum mechanics of unsharp observables (POV-measures) was developed recently in detail and is now well established. In particular, the quantum theory of measurements was elaborated for the more general case of unsharp observables.[25] Unsharp observables are understood here as unsharp properties that pertain objectively to a quantum system. It is not meant here that the value of an observable is merely subjectively unknown but objectively decided. Instead, we are faced here with an objective un-decidedness of properties.

For the relaxation of the ontology $O(Q)$ in the sense of unsharp properties we refer to the brief remarks in Section 1.3. Accordingly, the new task is the formulation of an unsharp quantum ontology, that will be denoted here by $O(Q^U)$. The

---

[21] Solèr (1995).

[22] Piron (1976) and MacLaren (1965).

[23] Busch et al. (1996).

[24] Mittelstaedt (1998).

[25] Busch et al. (1996).

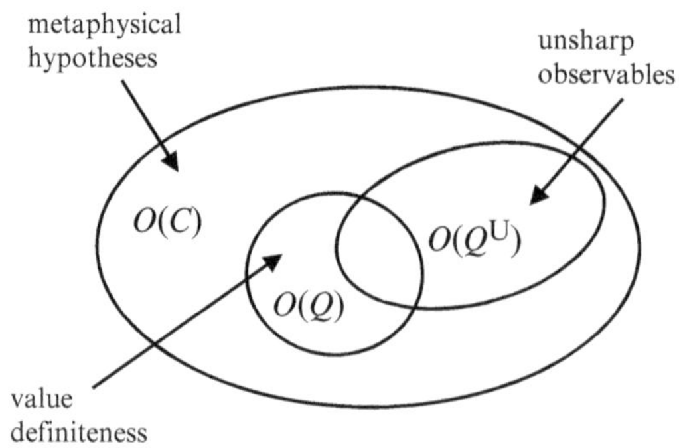

Fig. 3.10 Comparison between the ontologies $O(C)$, $O(Q)$ and $O(Q^U)$

quantum ontology $O(Q)$, which is characterised by the requirements $O(Q)^1 \ldots O(Q)^4$ in Section 3.2 is not yet in complete accordance with quantum physics for the following two reasons. First, the most general observables in quantum mechanics – the POV-measures – correspond to unsharp properties that allow for unsharp joint properties, even for complementary observables. Hence, the ontology $O(Q)$ is too restrictive since, generally, it does not allow for joint properties of complementary observables. Second, the requirement of value definiteness for all properties cannot be fulfilled, since the pointer objectification in the measurement process cannot be achieved in general. Hence, in this respect, the ontology is not sufficiently restrictive.

The two objections against $O(Q)$ can both be taken into account if $O(Q)$ is replaced by the new quantum ontology $O(Q^U)$ of unsharp properties, provided the degree of unsharpness is conveniently defined. It is a difficult question, how much unsharpness is needed quantitatively for removing the two deficiencies of $O(Q)$ mentioned. In Section 4.3, where we investigate the meaning of Planck's constant in quantum physics, we can provide an answer to this question.

Comparing the three ontologies in question, we find that on the one hand, $O(Q^U)$ is partly stronger than $O(Q)$ since it allows for unsharp joint complementary properties that are not contained in $O(Q)$. On the other hand, $O(Q^U)$ is partly weaker than $O(Q)$ since value definiteness of properties is not required. However, $O(Q^U)$ is weaker than the classical ontology O(C). These relations are illustrated in Fig. 3.10.

From the logical point of view, one has to show that the ontology of unsharp properties leads to a language and logic of *unsharp propositions*. At this point a clear distinction must be made between two different relaxations of the concept of value definiteness. For a sharp (s) proposition A value definiteness means that either A or the counter proposition $\bar{A}$ can shown to be true. In other words, there is a proof procedure that leads either to A or to $\bar{A}$. In intuitionistic logic L_i and in intuitionistic quantum logic L_{Qi} this value definiteness (v) is not assumed but replaced by the relaxed value definiteness [v] which means that in general neither a proof of A nor a proof of $\bar{A}$ is known. However, if accidentally a subset $\{A_1 \ldots A_n\}$ of elementary propositions is value definite in the sense of (v), then this value definiteness is inherited to all propositions which are finitely connected by the A_i and the logical connectives $\wedge$, $\vee$, $\neg$, and $\rightarrow$.

If sharp (s) propositions are replaced by propositions which are in general unsharp, expressed by $\{s\}$, then for an unsharp proposition A the two possibilities A and $\bar{A}$ are no longer mutually exclusive. This means that the properties $P(A)$ and $P(\bar{A})$ can both be attributed to the object system S. Under these conditions, the proposition $A \wedge \bar{A}$ is no longer formally false, which leads to various kinds of paraconsistent logical systems. The relaxed value definiteness that is induced by the unsharpness assumption $\{s\}$ will be denoted here by $\{v\}$. This weakened value definiteness means that the unsharp predicates A and $\bar{A}$ will both objectively pertain to the system. This corresponds to a realizable experimental situation in a measurement process. Hence we find that there are two quite different relaxations of the value definiteness of orthomodular logic:

(i) intuitionistic value indefiniteness $[v]$
(ii) unsharp value indefiniteness $\{v\}$

In unsharp quantum logic $L_{Qi}{}^u$ both relaxations are taken together. This means that even sharp propositions are value indefinite in the sense of $[v]$

In the quantum logic L_{Qi} of sharp propositions we made use of a relaxation of the availability (a) which is replaced by the restricted availability $[a]$. If we proceed from L_{Qi} to a quantum logic of unsharp propositions the relaxation of the availability that leads to condition $[a]$ is too strong and must be partially retracted. Indeed, the amount of availability increases with the unsharpness of the propositions in question. If after the successful proof of the unsharp proposition A by a measurement another unsharp proposition B is proved, then the result of the previous A-measurement is partially still available. This can be demonstrated experimentally as well as theoretically within the context of a simple Mach-Zehnder interference experiment.[26] Note however, that the conveniently defined amounts $u(A)$ and $u(B)$ of unsharpness of A and B, respectively, must fulfil the Heisenberg relation $u(A)u(B) \geq \hbar/2$.[27] This partially retracted, restricted availability will be denoted here by $\{a\}$.

This means that the new and modified availability is restricted via uncertainty relation by Planck's constant. In other words, if the mutual uncertainties (i.e. the product $u(A)u(B)$) remain below $\hbar$ then we fall back to the rigorously restricted availability $[a]$. – We summarize the various pragmatic preconditions of elementary (material) propositions in the following table.

[26] Mittelstaedt et al. (1987)

[27] Busch (1985) and Busch et al. (2007).

(s)		{s}
(v)	[v]	{v}
(a)	[a]	{a}

Properties of elementary propositions

The logical calculi are related to these properties in the following way:

(s): (v) (a)	L_C
[v] (a)	L_i
(v) [a]	L_Q
[v] [a]	L_{Qi}
{s}: {v} {a}	L_{Qu}

These considerations show that elementary (material) propositions are in general characterised by the unsharp value definiteness $\{v\}$ and by the relaxed restricted availability $\{a\}$. The new task is then the explicit construction of a formal logic of unsharp propositions in the sense of $\{s\}$ by means of a dialog game, say. This has not yet been done except one not completely satisfying attempt.[28] On the basis of $\{s\}$, $\{v\}$ and $\{a\}$ we expect an unsharp (fuzzy) paraconsistent logic with commensurability restrictions that are weaker than the restrictions in intuitionistic and orthomodular quantum logics L_{Qi} and L_Q. In the literature, we find a large variety of proposals for unsharp quantum logics. We mention here: unsharp quantum logic,[29] effect algebras and unsharp quantum logics,[30] Brouwer-Zadeh quantum logic,[31] fuzzy intuitionistic quantum logic,[32] partial unsharp quantum logic.[33]

Presently, it is not known, whether one of these logical systems corresponds to the requirements formulated in this chapter, i.e. whether it can be reconstructed by the operational techniques mentioned. Hence, the proposed $\{v\}$-$\{a\}$-dialog semantics could serve as a criterion, as a guiding principle for finding the "true" unsharp quantum logic. The new task is rather ambitious. In a first step, it must be shown that the generalised dialog semantics leads – if at all – to one of the proposed logical systems. With respect to the dialog-semantics, soundness and completeness must be proved. In the proposals mentioned soundness and completeness were proved merely with respect to an algebraic semantics. This is, however, not

[28] Giuntini and Greuling (1989).

[29] Dalla Chiara (1995).

[30] Foulis and Bennett (1994).

[31] Giuntini (1990).

[32] Dalla Chiara et al. (1993).

[33] Dalla Chiara and Giuntini (1994).

sufficient for demonstrating consistency between logic and the underlying ontology that is expressed by the pragmatic preconditions of the formal language in question.

In the revised approach to quantum logic, which was discussed in the preceding Section 3.4, we assumed that elementary (material) propositions correspond to unsharp properties in the sense of POV-measures. The formal logic of these "unsharp propositions" leads to a Lindenbaum-Tarski algebra that will – presumably – agree with one of the proposed algebras of unsharp properties. We expect that in a final step this way of reasoning leads to the effect algebra in Hilbert space and to quantum mechanics of unsharp properties in Hilbert space. For the ontological preconditions of quantum logic this means that the assumed properties of elementary (material) propositions should be reproducible as outcomes of a preparatory quantum mechanical measurement process. This is the requirement of consistency between quantum logic and the underlying ontology mentioned above. However, even for the unsharp propositions discussed in the last section, this requirement cannot be fulfilled.

As mentioned above (Section 3.4) the first attempt to solve the measurement problem consists in generalising the concept of object-observables by replacing sharp observables (projection valued measures) by unsharp observables that are given by POV-measures (positive operator valued measures). This attempt was based on the conjecture that the quantum measurement problem is induced by the use of idealised sharp observables. This is, however, not the case. It could be shown that not only for sharp observables but also for unsharp object observables in the sense of POV-measures, pointer objectification cannot be achieved.[34] Moreover, we do not know up to now, whether a stronger relaxation of the ontological premises can solve the measurement problem.[35]

However, this negative result does not invalidate directly the generalisation of orthomodular quantum logic, which we investigated in the present section. The generalised quantum logic, which is based on unsharp elementary propositions in the sense of POV-measures, presupposes more adequate ontological hypotheses than orthomodular quantum logic. Hence, this new logic, which is again weaker than classical logic is applicable to a domain of physical phenomena that is more in accordance with quantum mechanics than orthomodular logic. It is not claimed here that quantum logic of unsharp propositions is free from ontological hypotheses at all and thus applicable to the totality of physical phenomena.

Quantum logic of unsharp propositions is not the "true logic". Since unsharp propositions do not solve the measurement problem, we cannot expect that unsharp quantum logic is the "final logic" of physics, which is in accordance with the universal "final theory of everything". However, unsharp quantum logic is closer to the "final logic" than orthomodular logic and classical logic. This means, in addition, that due to the more adequate relaxation of non-empirical ontological hypotheses quantum logic of unsharp properties is more intuitive and more plausible than quantum logic of sharp properties and classical logic.

---

[34] Busch and Shimony (1996).

[35] Busch (1998).

3.6 Statistics of Physical Objects

3.6.1 Natanson Statistics

The rational reconstruction of quantum mechanics which we discussed in the present Chapter 3, was guided by the idea that the essential features of quantum mechanics can be obtained from classical mechanics merely by abandoning or relaxing the metaphysical hypotheses $O(C)^1 \ldots O(C)^6$ contained in the classical theory. For the reconstruction of quantum mechanics, more precisely of non-relativistic quantum mechanics, we had to abandon only the hypotheses $O(C)^3 \ldots O(C)^6$, whereas the hypotheses $O(C)^1$ and $O(C)^2$ of the existence of an absolute time and of the validity of the Euclidean geometry in a relative space, respectively, could be preserved without any change. Presently, it is not known whether the elimination of these two hypotheses would lead to an improvement of the already reconstructed quantum mechanics.

However, it would be worthwhile to know how the elimination of the hypotheses $O(C)^3 \ldots O(C)^6$ is expressed in detail in the reconstructed quantum mechanics. Generally, it is rather hard to see which feature of the reconstructed quantum mechanics is induced by the elimination of which hypothesis. The reconstructed full theory is the result of abandoning or relaxing the totality $O(C)^3 \ldots O(C)^6$ of hypotheses without thereby indicating the influence of a particular hypothesis. There is, however, a remarkable and illuminating exception. If we eliminate or relax in classical mechanics nothing, except the hypothesis $O(C)^3$ which states that there are individual objects, then we obtain a classical theory with objects that are indistinguishable. Except from this conceptual deficiency, the reconstructed theory would not show any remarkable quantum features.

The only difference between this pseudo-classical theory and the true classical mechanics is that the statistics of bodies is no longer governed by Boltzmann statistics. Indeed, the statistics of indistinguishable objects, which was first discovered in the statistical behaviour of photons by Planck and Einstein, was later elaborated generally as the statistics of indistinguishable objects by Natanson[36] and finally applied to photons and massive particles by Bose[37] and Einstein.[38] Today, it is called Bose-Einstein statistics. In order to illustrate this result we consider, according to Natanson, P indistinguishable particles and N distinguishable boxes of finite size and investigate the distribution of P particles over N boxes. In case of classical, i.e. distinguishable particles there are

$$\mu_C(P,N) = N^P$$

[36] Natanson (1911).

[37] Bose (1924a, b).

[38] Einstein (1924, 1925).

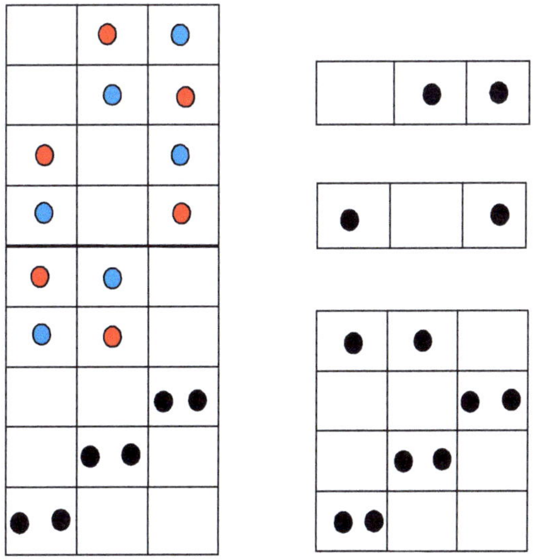

Fig. 3.11 Comparison of classical statistics and Natanson statistics

ways to distribute the P particles over N boxes. For indistinguishable particles there are

$$\mu_{Na}(P,N) = \frac{(N+P-1)!}{P!(N-1)!}$$

possible ways for distributing the particles over the N boxes. This is the kind of distribution, which is known today as *Bose-Einstein statistics*. Here, however, we will not apply this terminology. Instead, we call it *Natanson statistics* for reasons that will be explained later. In contrast to the classical situation, in Natanson statistics permutations of particles in a box do not lead to new cases. Hence, a large number of configurations, which in the classical case are considered different, are counted as one in Natanson statistics. For a more intuitive illustration, we show in Fig. 3.11. the possible ways to distribute two particles ($P = 2$) over three boxes ($N = 3$) for classical statistics according to $\mu_C(P, N)$ as well as for Natanson statistics according to $\mu_{Na}(P, N)$.[39]

The physical relevance of the Natanson statistics becomes clear if we identify the boxes with states in the sense of quantum mechanics. In the limit of high temperature, the number of possible states is increasing and the number of configurations with two or more particles in one state becomes negligible. The difference to classical statistics disappears in this case, except from a simple reduction of the statistical weight of any configuration by a factor $P!$ compared with the classical case. By contrast, for low temperature the number of configurations with two or more particles occupying the

[39] Adopted from Hund (1967), p. 154.

same state is not negligible, and those configurations are privileged. This means that at low temperature Bose-particles have a greater probability than classical particles to occupy the ground state. This phenomenon is known as Bose-Einstein condensation.

As already announced above, we will explain now somewhat more in detail our change of the terminology (replacing "Bose statistics" by "Natanson statistics"). The Bose-Einstein statistics was obtained originally by physical considerations about the properties of photons and of massive particles with integer values of the intrinsic spin, the massive "bosons". Some of these physical phenomena were mentioned in the preceding paragraph, e.g. the Bose-Einstein condensation. By contrast, we obtained the statistical distribution according to Natanson by using the indistinguishability of objects as a new and weaker ontological hypothesis, which replaces the stronger hypothesis $O(C)^3$. For this reason, we call here the new statistics "Natanson statistics". In addition, we mention here that from a physical point of view, we know presently two different kinds of elementary particles. Particles with integer values of the intrinsic spin are called bosons since they are subject to the Bose statistics, which we denoted here as Natanson statistics. Particles with half-integer values of the intrinsic spin are called fermions since they are governed by the Fermi statistics, which will be discussed in the next Section 3.6.2. The connection between these two kinds of particles and the corresponding two statistics is established by the spin-statistics theorem: Bosons are not governed by Fermi statistics and fermions are not governed by Bose statistics. W. Pauli gave the proof of this fundamental theorem[40] within the framework of relativistic quantum field theory in 1940. The proof is difficult and cannot be presented here. For details we refer to the literature.[41]

Irrespective of these problems, in the present chapter there is no need for a detailed treatment of this rather ambitious theorem. Indeed, we obtained the Natanson statistics exclusively by abandoning the ontological hypothesis $O(C)^3$ in the classical ontology $O(C)$. Unfortunately, a similar way of reasoning for the Fermi statistics in the next Section 3.6.2 is not in sight.

3.6.2 *Leibniz Statistics*

We consider again the ontological hypothesis $O(C)^3$ of the classical ontology but for pedalogical reasons we decompose it here into three independent requirements.

(a) There are individual and distinguishable objects
(b) These objects can be named and identified at a certain instant t of time and re-identified at a later time $t > t'$.
(c) The number n of objects with certain properties can assume any finite integer value n.

---

[40] Pauli (1940).

[41] Streater and Wightman (1964), Section 4.4.

As we have shown in Section 3.6.1. the abandonment of the requirements (a) and (b) leads to a possible world whose objects are indistinguishable. From a statistical point of view, these objects are subject to the Natanson statistics. However, even under these restricted conditions we have tacitly assumed what is formulated here as requirement (c). It is obvious that this residual assumption can further be specified, e.g. by restricting the free choice such that a certain complete set of essential and accidental properties can pertain only to one object system, say. In physics, and in particular in atomic physics, this requirement is usually called "Pauli principle". Furthermore, with respect to the statistics it is useful to distinguish two concepts of indistinguishablility: First, the genuine indistinguishability that refers to the outological identity of the object in question and second the epirtemic indistinguishability that refers to the empirical impossibility to distinguish the object considered from other objects.

This would mean that two objects that agree with respect to all properties (essential and accidental) are identical. Conversely, it means that there are no two different objects which agree in all essential and accidental properties. Similar as in the case of Natanson statistics, we make use here of a mere abstract and conceptual characterisation of the Leibniz statistics mentioned above for deriving the corresponding statistics. This way of reasoning allows for a comparison of the statistics according to Boltzmann, Natanson, and Leibniz, as illustrated in Fig. 3.12. In the philosophical literature, this statement is well known as the *Principium Identitatis Indiscernibilium*, which was formulated in this way first by Leibniz and repeated on many occasions; *'il n'est pas vray que deux substances se ressemblent entierement, et soyent differentes solo numero.*[42] Here we call the statement "Leibniz principle" and denote it by the acronym "*PII*" for *Principium Identitatis Indiscernibilium.*

This principle has a long history. Leibniz adopted the *PII* from Thomas Aquinas, who had demonstrated the validity of this principle for angels. According to Thomas, angels are spiritual, simple and indivisible beings that are completely determined by their essential properties. Hence, every angel is unique and there are as many kinds of angels as there are particular angels.[43] Leibniz refers explicitly to the writings of Thomas Aquinas and asserts, "What S. Thomas assures about the angels must be considered to be valid for all substances"

> *que ce que S. Thomas asseure sur ce point des anges ou intelligences* (quod ibi omne individuum sit species infima) *est vray de toutes les substances, ...*[44]

We should repeat that in physics elementary particles with half integer values of the intrinsic spin are called "fermions" and elementary particles with integer values of the intrinsic spin are called "bosons". However, we emphasise again that these physical circumstances are not the basis of the present considerations. Here we derived the Natanson statistics from the ontological assumption of the indistinguishability of the objects. In addition, we will show that objects that are also

[42] Leibniz, G. W. Discours de Métaphysique, GP IV, p. 433.

[43] Castellani and Mittelstaedt (2000).

[44] Leibniz, G. W., Discours de Métaphysique, GP IV, p. 434.

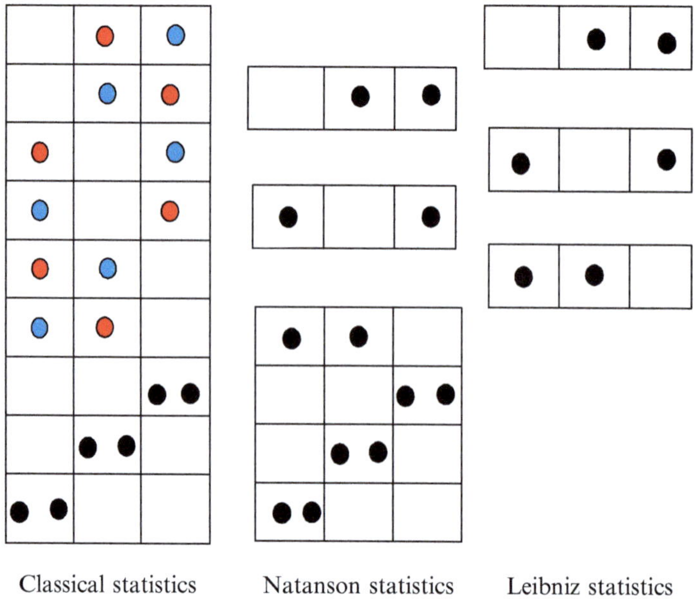

Fig. 3.12 Comparison of classical statistics, Natanson statistics and Leibniz statistics

subject to the Leibniz principle are governed by another statistics, which will be called here – for obvious reasons – "Leibniz statistics". Although this Leibniz statistics agrees formally with the Fermi statistics mentioned above in Section (3.6.1), we did not refer to these physical results at all. In particular, our present derivation shows that similarly as in the case of bosons also for fermions we reconstructed a pseudo-classical theory which does not show any remarkable quantum features.

In order to illustrate the Leibniz statistics more formally, we consider again P particles and N single particle boxes. In the case of the Leibniz principle there are $\mu_{PII}(P,N)$ ways to distribute P particles over N boxes with $\mu_{PII}(P,N) = \frac{N!}{P!(N-P)!}$. Hence for $P = 2$ particles and $N = 3$ boxes we have $\mu_{PII}(2,3) = 3$ ways to distribute the particles over the boxes (Fig. 3.12).[45]

For illustrating the metaphysical *principium identitatis indiscernibilium* by a convenient everyday example, Leibniz describes how an aristocratic circle vainly tries to find two perfectly similar leafs in the garden of Herrenhausen: ... *"une grande Princesse, qui est d'un esprit sublime, dit un jour en se promenant dans son jardin, qu'elle ne croyoit pas, qu'il y avoit deux feuilles parfaitement semblables. Un gentilhomme d'esprit, qui estoit de la promenade, crût qu'il seroit facile d'en trouver; mais quoyqu'il en cherchât beaucoup, il fut convaincu par ses yeux, qu'on pouvoit tousjours y remarquer de la difference"*[46]

[45] Adopted from Hund (1967), p. 154.

[46] G. W. Leibniz, GP V, p. 214. – (Engraving reproduction; Schubert 1796).

For Leibniz, the significance of the *PII* is very high. In his letter of 02.06.1716 to Clarke, he tried to convince Clarke (and in this way also Newton) of the importance of this and other metaphysical principles: "*Ces grand principes de la raison suffisante et de l'identité des indiscernable, changent l'etat de la Metaphysique, qui devient reelle et demonstrative par leur moyen : au lieu qu'autres fois elle ne consistoit presque qu'en termes vuides.*"

"*Poser deux choses indiscernables, est poser la même chose sous deux noms*".[47]

3.7 Are the Laws of Quantum Logic Laws of Nature?

The reconstruction of quantum mechanics by abandoning several hypothetical assumptions of classical mechanics allows one also to give a – perhaps preliminary – answer to the difficult and ambitious question, whether the laws of quantum logic

[47] G. W. Leibniz, GP VII, p. 372.

are genuine laws of nature. Without going too much into details, already at this point, by a law of nature we understand a contingent, albeit universally valid law that refers to the external material reality. Here, we do not think of the huge number of rules that are often denoted as "laws", like Ohm's law, but of the most fundamental law-like structures in physics as *classical mechanics, quantum mechanics, electrodynamics, general relativity, etc.* For more details we refer to the literature.[48]

The early history of quantum logic provides a rather confusing picture. In his pioneering book of 1932, J. von Neumann[49] made the observation, that the projection operators in a Hilbert space may be considered as elementary yes – no propositions about measurable properties and he constructed also the most important logical connectives "and", "or", and "not" in terms of projection operators. In a subsequent publication with G. Birkhoff of 1936,[50] entitled *The Logic of Quantum Mechanics* the authors could show that the projection operators and the corresponding propositions constitute an orthocomplemented lattice with some additional properties. In contrast to the orthocomplemented and distributive (Boolean) lattice of classical logic, the lattice of "the logic of quantum mechanics" turned out to be much weaker than the Boolean lattices being neither distributive nor modular.

However, this formally well elaborated first version of "quantum logic" was not generally accepted by the scientific community of physicists and philosophers. The critique against "the logic of quantum mechanics"[51] was mainly concerned with the objection, that "the logic of quantum mechanics" is not a genuine logic in the strict sense, a formal structure in the tradition of Aristotle, Thomas Aquinas, and George Boole, which governs the rules of our rational thinking and arguing. The reason for this refusing reaction to quantum logic was the suspicion that either simple mistakes were made in the presentation of quantum logic,[52] – or that the advocates of quantum logic tacitly ignored the fact that the deviations of quantum logic from ordinary classical logic are induced by empirical results of quantum mechanics. Hence, quantum logic could also be obtained in a very natural way, namely, as Putnam wrote, just by reading "the logic off from Hilbert space".[53] If these assumptions were true, then the laws of quantum logic would indeed be laws of nature in a somewhat disguised form. – We add that in the sceptical contributions to quantum logic, we do not find any critical remark about the justification of classical logic. Instead, classical logic was taken for granted and only the deviations from this "true" logic were subject to the critique mentioned.

The operational approach to quantum logic, which makes use of a reduction or relaxation of metaphysical hypotheses of classical ontology in establishing a formal

[48] Armstrong (1983), Mittelstaedt and Weingartner (2004) and Van Fraassen (1989).

[49] von Neumann (1932).

[50] Birkhoff and von Neumann (1936).

[51] For a very detailed presentation of the history of quantum logic cf. Jammer (1974), pp. 341–616.

[52] Feyerabend (1965) and Stegmüller (1970), pp. 438–461.

[53] Putnam (1969).

language and logic of quantum physics, was only gradually accepted by the scientific community since the early 1960s. In this approach, which is presented here in Sections 3.3 and 3.4, quantum logic appears as an a priori structure that is justified more rigorously and under weaker assumptions than the laws of classical logic. This means that first of all the pretended preference of classical logic is no longer justifiable and secondly, that quantum logic contains fewer empirical contributions – if at all – than classical logic. In other words, at this stage of our discussion, there is no need to care about possible empirical components in quantum logic. Hence, our first, still preliminary result is that the laws of quantum logic are a priori true and should not be considered as laws of nature.

On the basis of the operational justification of quantum logic which refers to a relaxed quantum ontology,[54] the a priori foundation of quantum logic can be made more explicit and the formal algebraic structure can be elaborated more in detail. (Cf. Section 3.4) In particular, it turns out, that the Lindenbaum-Tarski algebras of the calculi of formal quantum logic are lattices. In classical logic, the lattice structure of the calculi of formal logic was already well known, e.g. from the investigations of Lorenzen[55] and Curry.[56] In a similar way, the calculi of intuitionistic quantum logic and of the full orthomodular quantum logic lead to corresponding lattice structures. These investigations confirm in some sense the above mentioned conjecture, that quantum logic in its various representations is a formal structure that follows a priori from the most general preconditions of a scientific language of physics. At this stage of our treatment, there are no law-like contributions in sight in the sense of laws of nature.

However, only on this very high formal level of considerations a completely new aspect becomes meaningful, which we called in Section 3.5 "the two faces of quantum logic". The orthomodular quantum logic, which we constructed in Section 3.4 is not only a consequence of the quantum ontology $O(Q)$, but presumably also its origin. Namely, once we have an orthomodular lattice that refers to predicates of a single system,[57] then it seems possible to proceed in a few steps to the lattices L_H of subspaces of Hilbert space. By means of the important theorems of Piron, Keller, and Solèr – mentioned in Section 3.5 – we finally arrive at quantum mechanics in Hilbert space $H(C)$ over the field of complex numbers C. There are two possible interpretations of this important result: The long way from quantum ontology $O(Q)$ via quantum language, quantum logic, quantum lattices to quantum mechanics in a Hilbert space $H(C)$ shows, that quantum mechanics itself is a priori true in the sense of a transcendental justification. Secondly, since quantum mechanics is usually considered as an empirical structure that corresponds to a genuine law of nature, also the laws of quantum logic, which finally lead to quantum mechanics must contain empirical components that at the end of this way of reasoning imply the

[54] Stachow (1976) and Mittelstaedt (1978).

[55] Lorenzen (1955).

[56] Curry (1952).

[57] As mentioned in Section (3.4) a lattice of this kind is atomic and fulfils the covering law.

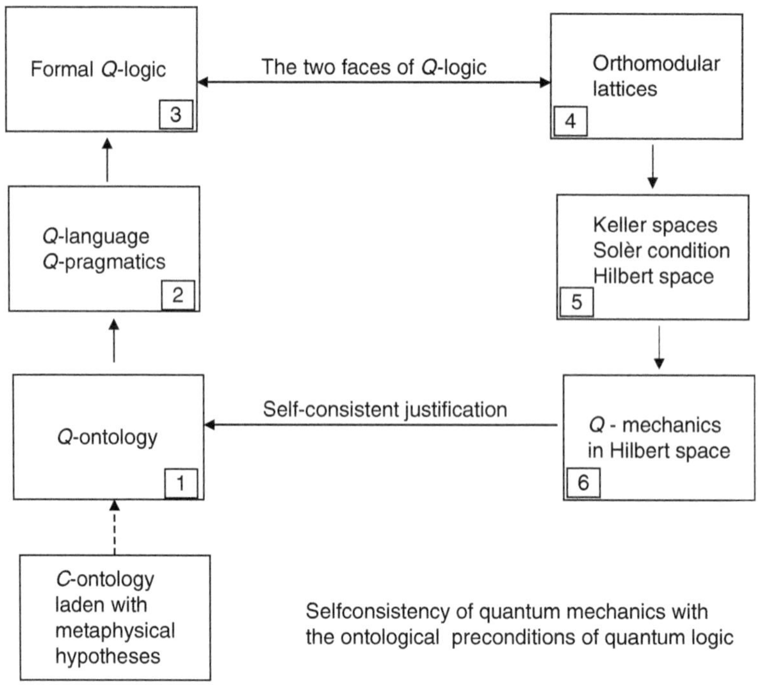

Fig. 3.13 Self-consistency between *Q*-ontology and quantum mechanics

empirical components in quantum mechanics. Where do these empirical elements in quantum mechanics come from?

The answer to this question is given by the theory itself. By means of a consistency argument – sometimes called "self-consistency" – the empirical components that are still contained in the quantum ontology $O(Q)$, and presupposed in the whole way of reasoning, can be justified by quantum mechanics in Hilbert space. Indeed, the remaining weak ontological premises $O(Q)^1 \ldots O(Q)^4$, which are free from metaphysical hypotheses, turn out to be compatible with quantum mechanics in Hilbert space. In this way, quantum mechanics can justify that ontology from which quantum logic and quantum mechanics evolved.

The whole way of reasoning is graphically represented in Fig. 3.13. The *Q*-ontology in box [1] contains an empirical physical component that is free from metaphysical contributions. The formal structures that arise from this quantum ontology appear as a priori structures, as long as the empirical physical component of $O(Q)$ is ignored. However, the various structures in this diagram, *Q*-language in box [2], *Q*-logic in box [3], *Q*-lattices in box [4], preserve the empirical physical component, which finally leads to *Q*-mechanics in Hilbert space in box [6]. And this final theory justifies the physical content in question in the initial quantum ontology in box [1] and confirms in this way the whole way of reasoning. The laws of quantum logic originate from certain most general features of nature in box [1] and they lead – after incorporating various mathematical structures – to the laws of nature in box [6]. However, in the present context, the laws of quantum logic are no laws of nature but a formal structure that evolves a priori from the preconditions of a scientific language of quantum physics. Obviously, the answer to the question, whether the laws of quantum logic are laws of nature, is strongly context dependent.

For an adequate understanding of these interrelations we have to explain the fact, that obviously the weak ontological preconditions formulated in box [1] are sufficient for deriving the full quantum mechanics in box [6]. This means that the empirical physical content of quantum mechanics is not larger than the empirical content in the initial ontology $O(Q)$ in box [1]. In this clearly defined sense, quantum ontology and quantum mechanics are equivalent. With respect to the empirical physical content, quantum ontology is sufficient and necessary for quantum mechanics. This implies that quantum mechanics in Hilbert space is a theory that does not contain much information about the empirical reality that we call nature. But obviously, the same theory contains a lot of information about our ways to grasp this physical reality by means a formal language, its syntax and semantics, and a formal logic, together with various mathematical structures. In the course of our reconstruction of quantum mechanics from the initial ontology in box [1], these linguistic and mathematical tools were added to the original quantum ontology. This result agrees with the well known observation that quantum mechanics is at bottom an empty theory, a formal framework that must be filled step by step with empirical content. Also quantum ontology is only a framework that must be filled with linguistic and mathematical structures. Hence, from a structural point of view, quantum mechanics in box [6] is much richer than the ontology in box [1]. However, quantum mechanics does not possess more empirical physical content than $O(Q)$ – but also not less if it confirms the ontology $O(Q)$ by a self-consistency argument.

At this stage of our consideration we must argue very carefully. The assumed quantum ontology, which is based on the requirements $O(Q)^1 \ldots O(Q)^4$, implies two important premises for establishing a language and logic of quantum physics, (a) the restricted availability of propositions and (b) the value definiteness and reliability of elementary propositions. Premise (a) can be confirmed by the self-consistent justification of the ontology $O(Q)$. However, the premise (b) is too strong. According to the quantum theory of measurement we can objectify only properties that are not value definite and – even in its unsharp version – not reliable. This inconsistency between quantum mechanics and quantum ontology, which we mentioned already in Section 3.5, is known as the "problem of objectification" or the "measurement problem".

It is, of course, an important question, how quantum mechanics can lead to results that are not compatible with the presupposed quantum ontology $O(Q)$. The reason is that we have tacitly assumed here that the correctly reconstructed quantum mechanics is in addition universally valid and applicable to all domains of the physical reality. This assumption, which was not justified here, implies that quantum mechanics can be applied also to the quantum mechanical measuring process. This means that quantum mechanics governs not only the phenomena of the microscopic quantum world, but also the partly macroscopic means of its own verification. In the original formulation of the quantum ontology *O*(*Q*), this goal was not envisaged at all, and thus the compatibility of the extended quantum mechanics with the original ontology was still an open question.

The present situation is quite similar to the situation in the early days of operational quantum logic, when the classical ontology *O*(*C*) was replaced by *O*(*Q*). Indeed, the ontological preconditions contained in *O*(*Q*) are too strong compared with the reconstructed quantum mechanics. Hence, self-consistency cannot be achieved. However, *O*(*Q*) is not loaded with metaphysical hypotheses but merely with simplifying assumptions. In order to re-establish consistency between quantum mechanics and the underlying ontology, *O*(*Q*) must be further relaxed, in particular with respect to the value definiteness of elementary properties. In Section 3.5 about the bottom-up reconstruction of quantum mechanics in Hilbert space, we discussed in detail the difficulties of this project which are not yet resolved. As the envisaged result of such efforts we expect self-consistency between quantum mechanics and a new weak ontology $O(Q^{\mathrm{U}})$ of unsharp properties.[58]

On the basis of these considerations, we will try to answer the question in the title of the present section. Under the optimistic assumption that self-consistency can be achieved by means of this new and weak unsharp ontology $O(Q^{\mathrm{U}})$, the empirical physical content of this ontology will be contained also in the laws of quantum logic. In this sense, the laws of unsharp quantum logic are based on empirical, physical results that might be considered as laws of nature, but the laws of unsharp quantum logic itself are not genuine laws of nature. We emphasise again that the empirical component mentioned is not a particular feature of ordinary (orthomodular) and of unsharp quantum logic, since the empirical, physical content of classical logic that originates from classical ontology *O*(*C*) is much larger. As mentioned already at the end of Section 3.5, for this reason unsharp quantum logic is closer to a hypothetical "final logic" than orthomodular logic and classical logic.

3.8 Quantum Physics and Classical Physics – Their Respective Roles

The reconstruction of quantum mechanics which we have presented here allows one to answer the question whether in this theory the ideas, the goals, and the philosophical interpretations of the founders of quantum mechanics are actually realised. In Bohr's approach, which we have briefly sketched in Section 3.1, as well as in the following two or three decades quantum mechanics was considered always from the perspective of classical physics. The apparatuses for the empirical justification were considered as objects of classical physics and the interpretation of the

[58] We will come back to this ontology $O(Q^U)$ in Sections (4.3.1) and (4.3.4).

theory was based on classical concepts. Hence, classical physics seemed to be a necessary requisite for the justification and interpretation of quantum mechanics.

In the present chapter we provided a (rational) reconstruction of quantum mechanics which is based on the weak ontologies $O(Q)$ or $O(Q^{\mathrm{U}})$, and on the formal languages that can be established on the basis of these ontologies. The relaxation of the classical ontology $O(C)$ consists of a relaxation or elimination of hypothetical assumptions that are contained in $O(C)$ but which can neither be justified by rational arguments nor by empirical evidence. The result of our reconstruction is a formal quantum mechanics in Hilbert space, which reflects merely the structures of the relaxed quantum ontology. This reconstructed quantum mechanics is an empty theory which must still be filled with material content. However, the theory is presumably universally valid and within the framework of the given ontology also not disprovable. As to the quantum ontology in question we note that it is neither completely free from hypothetical assumptions, nor is it obvious that it can grasp all relevant structures of the quantum world. Hence, on the level of the ontology, there is still room for improvement and revision.

This new foundation and justification of quantum mechanics shows that many of the pretended strange features of the theory have nothing to do with the theory but at most with the ways of its discovery. As in classical physics also in quantum mechanics objects are described by their properties but with the difference that objects are not thoroughgoing determined. However, this restriction has nothing to do with the measurement apparatuses. On account of the universal validity of quantum mechanics also the apparatuses are quantum objects and governed by the restricted quantum ontology. In the quantum theory of measurement an apparatus is part of the quantum world and its connection with object systems is described by the same theory as entanglement.

The interrelation between quantum mechanics and classical physics now becomes obvious. Quantum mechanics is based on a weaker ontology than classical physics and depends, for this reason, on less not sufficiently justified hypotheses. The theory is valid in the large domain of reality that corresponds to the weak ontology. In particular, quantum theory holds for the measurement apparatuses and that irrespective of the question whether they are – according to the traditional view – considered as microscopic or macroscopic entities. This means that for measurement processes and for the interpretation of quantum mechanics classical physics is completely dispensable. Classical physics is not the methodological *a priori* of quantum physics. Not only with respect to the empirical justification but also with respect to its interpretation quantum mechanics remains with itself.

Chapter 4
Three Constants of Nature

4.1 The Problem of Constants of Nature in Modern Physics

In accordance with the results of the preceding Chapters 2 and 3, we assume that the two fundamental theories of modern physics, the theory of relativity (*SR*) and quantum mechanics (*QM*), can be obtained from the classical space-time theory and classical mechanics (*CM*) by abandoning or relaxing hypothetical assumptions contained in these classical structures. In addition, it became also obvious in the preceding chapters that classical mechanics and Newton's space-time theory describe a fictitious world that does not exist in reality. Hence, we do not expect to find in modern physics any indications that refer to classical mechanics and to Newton's space-time. Possible traces of absolute time and absolute space are equally eliminated in special and general relativity as the classical limit in quantum mechanics. In modern physics we can completely dispense with the pretended classical roots. In quantum mechanics as well as in relativity there is no need and no room for anything like a correspondence principle or a non-relativistic limit, respectively.

Independent of these considerations and irrespective of their possibly far reaching implications, classical mechanics and classical space-time are often considered under a completely different angle as limiting cases of the theories of modern physics. According to this assessment the borderline between relativity and classical physics is characterised by the constant c, whereas the borderline between quantum mechanics and classical physics is determined by Planck's constant $\hbar$. Both quantities, c and $\hbar$ are thereby considered as fundamental constants of nature. However, in the light of the above mentioned arguments about the interrelations between Newton's classical physics and the fundamental theories of modern physics, it is rather hard to understand why the well established theories of modern physics should contain constants of nature that characterise the borderlines to a theory of a fictitious world. In other words, how can well established theories contain parameters which relate these theories to a theory of a fictitious world? Since there is no satisfying and convincing answer to this question, we will search in the

P. Mittelstaedt, *Rational Reconstructions of Modern Physics*,
Fundamental Theories of Physics 174, DOI 10.1007/978-94-007-5593-2_4,

following sections for alternative meanings of the constants c and $\hbar$. In particular, we will ask whether there is an intrinsic meaning of c within the theory of relativity and that without any relation to classical physics – and we will ask whether there is an intrinsic meaning of "$\hbar$" in quantum mechanics and that again without any recourse to classical mechanics.

4.2 The Meaning of the Constant c in Special Relativity

4.2.1 Preliminary Remarks

From a historical point of view, the theory of special relativity (*SR*) was understood at first as improvement and generalisation of classical physics that holds in the domains of very large velocities. As long as the new theory was considered as restricted to these domains of the physical reality, the boundary line between the new theory (*SR*) and the classical Newtonian theories of space-time could be characterised by the constant c (velocity of light). This understanding of the constant c is well known from the literature and can be found in many monographs and textbooks. However, in the first two or three decades of the twentieth century it became more and more obvious that the new theory of special relativity (*SR*) is not restricted to the relativistic domain mentioned but, compared with classical physics, is universally valid and applicable to all domains of the physical reality.[1] Hence, the interpretation of the constant c as indicating the boundary between classical physics and relativistic physics became meaningless. Indeed, on the basis of an increasing experimental accuracy of measurements it turned out that (*SR*) is also valid for motions that are slow compared with the velocity c of light. This implies that also another interpretation of the constant c is possible, the limiting case interpretation, which states that in the limit ($v/c \to 0$) of very slow motions compared with the velocity of light, special relativity (*SR*) approaches the physics of classical space-time. It is obvious, that the less careful formulation which replaces the limit mentioned above by the limit ($c \to \infty$) is very misleading since c is a constant with a well known numerical value of $3 \cdot 10^8$ m/s and will never approach infinity.[2]

If special relativity were interpreted and justified operationally by means of light rays, such that measurements of space-time intervals are generally performed by means of light signals, then the value of the velocity of light would have a direct and constituting influence on the observable space-time structure of (*SR*). In the first two or three decades of the twentieth century, Einstein and other scientists argued

---

[1] Except, of course of domains that are not considered in the present book.

[2] The first measurement of the velocity of light by O. Rømer was performed in 1670, several years before Newton's *Principia* appeared in 1686 (1st ed.) and 1723 (2nd ed.).

very often in favour of an operational interpretation of (*SR*). Many of the characteristic effects of (*SR*) were explained "intuitively" by measurements with light signals, i.e. traced back to the properties of light rays. We mention here in particular the synchronisation of distant clocks by light signals, the demonstration of the time delay of a moving clock and of the clock paradox by permanent communication between different observers with the help of light signals, and the theoretical construction of a light clock.[3] Only the attempt to explain also the phenomenon of the "Lorentz contraction" by means of optical observations[4] provided some problems. Indeed, R. Penrose[5] and J. Terrell[6] could demonstrate that the relativistic length contraction of moving bodies is not directly visible. However, irrespective of these critical remarks for at least 50 years the operational approach to special relativity by means of light rays was considered by the scientific community of physicists to be well established. We will not follow here this operational interpretation of special relativity further on, but replace it by a more realistic approach to space-time for reasons that are explained in the following sections.[7]

4.2.2 Metaphysics and Ontology

In contrast to his often quoted statement,[8] Newton's mechanics is based on many hypotheses that can be traced back to the metaphysics and theology of the seventeenth century, which are however, hidden and not made explicit in his work. In addition, for 200 years physics was based on these hypotheses, which were not questioned by physicists and philosophers. Only at the end of the nineteenth century, some of the metaphysical hypotheses of physics were exposed and made explicit, in particular by Mach, Poincaré, and somewhat later by Einstein. The progress of physics in the twentieth century is essentially the result of a stepwise reduction of metaphysical prejudices that were inherently contained in Newton's conception of physics. Here we are interested only in the first step of this long lasting historical process, the elimination of the concept of absolute time, which was introduced by Newton in his *Principia* with the words.

> Absolute, true and mathematical time, of itself and from its own nature flows equably without relation to anything external.[9]

---

[3] Martzke and Wheeler (1964) and Misner et al. (1973), p. 397.

[4] Cf. for instance Gamov (1946).

[5] Penrose (1959).

[6] Terrell (1959).

[7] For more details cf. Mittelstaedt (2006), p. 260.

[8] Hypotheses non fingo (Principia, 3. ed, p. 943).

[9] The original Latin formulation reads: "Tempus absolutum, verum et mathematicum, in se per natura sua absque relatione ad externum quodvis, aequabiliter fluit". Newton (1934), p. 6.

In Newton's writings we cannot find any indication of a rational or empirical justification of this statement. It is of a purely hypothetical character.

In order to elucidate the meaning of the deconstruction of absolute time in modern physics we start from the well known classical ontology *O*(*C*) of Newtonian physics, but without the assumption that there is an absolute and universal time. On the basis of this reduced ontology, which we denote here by *O*(*SR*), we will reconstruct a new, reduced physical theory of space-time. Obviously, this reduced ontology *O*(*SR*) will grasp a larger domain of reality than the old classical ontology *O*(*C*) since it is based on less postulates than Newton's classical ontology. Consequently, we expect that the physical theory of space-time that is reconstructed on the basis of the reduced ontology *O*(*SR*) is applicable to a larger domain of reality than classical physics and universally valid within the framework of phenomena discussed in the present book. In particular, we expect that this approach will provide new insights and perspectives of the main problem of the present investigation, the meaning of the constant c.

4.2.3 Reconstruction of Special Relativity

The reconstruction of special relativity that we have in mind here does not make use of any recourse to the propagation of light and its velocity. It follows essentially the detailed way of reasoning of the reconstruction of special relativity presented in Sections 4.2 and 4.3. As already mentioned in Section 4.1, this approach has a long history and can be traced back to the early days of (*SR*) in the beginning of the twentieth century.[10] We will not repeat the historical, but now somewhat antiquated arguments here, but instead refer to more recent results[11] and in particular the derivation of special relativity in Sections 4.2 and 4.3.

According to the general remarks in the last section we start with the ontology *O*(*SR*), which can be obtained from Newton's classical ontology *O*(*C*) if we dispense with the assumption that there exists an absolute and universal time. However, even under these restricted conditions we can start defining the concept of an inertial system. In a first step we introduce as in Section 4.2 an ensemble of small bodies $\Gamma = \{k_1, k_2, \ldots k_n\}$ which were freely thrust into the empty space. An inertial system can then be defined as a material basis of an observer that is equipped with rods and clocks for measurements of space-time intervals. A frame of reference of this kind is called an "inertial" system I if the elements of the constituting ensemble Γ move on straight lines in the sense of Euclidean geometry.[12] Since empirically

---

[10] Ignatowski (1910) and Franck and Rothe (1911).

[11] Levy-Leblond (1976) and Mittelstaedt (1976).

[12] It is an important and difficult problem how many test bodies must at least be used in the ensemble Γ in order to allow for deriving the linearity of the transformations between inertial systems. We will not discuss this question here and refer to the literature. Cf. Borchers and Hegerfeldt (1972).

the relative velocity of different bodies k_i and k_j does not depend on time, we can define a "metric time" by the requirement that the elements $k_n \in \Gamma$ are moving not only on straight lines but also uniformly, i.e. with time-independent velocities.[13]

An inertial system I can be equipped with a system $K_I(x_i, t)$ of coordinates of space and time, where x_i are Cartesian coordinates, say, of an Euclidean space and "t" the metric time as defined above. The transformation $T_{II'}$ that leads from coordinates $K_I(x_i, t)$ of I to the coordinates $K_{I'}(x'_i, t')$ of another inertial system I' transforms straight lines of the 4-dimensional (x_i, t)-space to straight lines of the (x'_i, t')-space. Hence, these transformations are at first collineations, which reduce, however, to linear transformations if we postulate that finite (x_i, t)-values are always transformed into finite (x'_i, t')-values. The linearity of the transformation $T_{II'}$ implies, that the velocity $v_{II'} = v$ of the system I' relative to the system I is constant with respect to time t.

In the following we consider, for the sake of simplicity only one spatial coordinate x. If we further assume that the systems of coordinates $K(x, t)$ and $K'(x', t')$ of I and I', respectively, coincide at $t = 0$, then the transformations $T_{II'}$ assume the simple form

$$x' = k(v)(x - vt), \quad t' = \mu(v)t + \nu(v)x$$

with three arbitrary functions $k(v)$, $\mu(v)$, and $\nu(v)$ that depend on the velocity v of I' relative to I. The following requirements are not substantially new postulates but consequences of the definition of the concept of an inertial system. Here we refer to the postulates 4, 5, and 6 of Section 2.2. Since the constituting ensemble Γ does not distinguish a special direction in space, we postulate isotropy of the transformation $T_{II'}$ (postulate 4). Since no system of inertia is distinguished with respect to Γ, we postulate the principle of relativity (postulate 5). Finally, we require that two inertial system I and I' are indistinguishable also with respect to a third inertial system I'' (postulate 6). These requirements imply the following restrictions for the functions $k(v)$, $\mu(v)$, and $\nu(v)$

$$k(v) = k(-v), \quad \mu(v) = \mu(-v), \quad \nu(v) = -\nu(-v).$$

Instead of the odd function $\nu(v)$ we will use here as in Section (2.2) the even function

$$\alpha(v) := -\nu(v)/v \cdot \mu(v).$$

If we introduce the notation

$$v := v_{II'}, \quad w := v_{I'I}$$

for the relative velocities of I and I', respectively, we obtain

[13] Mittelstaedt (1995), pp. 22–23.

$$v = v_{\mathrm{II}'} = -v_{\mathrm{I}'\mathrm{I}} = -w.$$

Making use again of the equivalence of $T_{I'I}\,(-v)$ and $T_{II'}{}^{-1}(v)$ we get

$$k(v) = \mu(v) = \left(1 - \alpha(v)v^2\right)^{-1/2}$$

and thus

$$x' = (x - vt)\left(1 - \alpha(v)v^2\right)^{-1/2}, \quad t' = (t - \alpha(v)vx)\left(1 - \alpha(v)v^2\right)^{-1/2}.$$

Hence, the remaining task is the determination of the unknown function $\alpha = \alpha(v)$. Using again the relativity principle and in particular the equivalence of two inertial systems I an I' with respect to a third one I'', we find that

$$\alpha(v) = \alpha(v') = \alpha = const.{>}0,$$

i.e. α is a positive constant that is independent of v.

Concluding this discussion we find that the constant α has a positive value, $\alpha = \omega^{-2} > 0$. Hence the transformation $T_{II'}\,(v, \omega)$ between two inertial systems I and I' with the relative velocity v reads (for one spatial coordinate x)

$$x' = \frac{x - vt}{\sqrt{1 - v^2/\omega^2}}, \quad t' = \frac{t - \frac{vx}{\omega^2}}{\sqrt{1 - v^2/\omega^2}}$$

and contains a positive constant $\omega^2 = 1/\alpha$, where ω is a velocity. At this point of our consideration we can already formulate two important results about the meaning of the constant ω.

1. Since in the case $\alpha > 0$ we have $-\,\omega \leq v \leq +\,\omega$ and find that the maximal velocity v between two inertial systems I and I' is given by ω.
2. Since the velocity v between two systems of inertia I and I' is restricted by ω, the temporal order of two events (E_1, E_2) in system I cannot in general be inverted by a transformation to system I'.

However, it should be emphasized here that in our derivation of the transformation $T_{II'}(v, \omega)$ there is no obvious reason to identify the invariant velocity constant ω with the velocity c of light. – Furthermore we note that the somewhat difficult decision for the case $\alpha > 0$ of a positive value of α is at the bottom a causality requirement since it can be traced back to the causal structure of the trajectories of the test bodies k_{i} of the constituting ensemble Γ, which structure should be preserved by a transformation between inertial systems.[14]

[14] For more details cf. Mittelstaedt (1995), pp. 83–116.

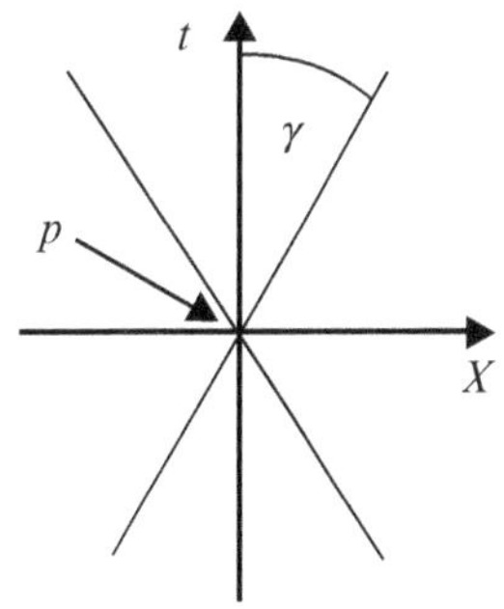

Fig. 4.1 The *null-cone* at point p and the cone angle γ

4.2.4 The Meaning of the Constant ω

The transformations $T_{II'}(v, \omega)$ between inertial systems I and I' – the inhomogeneous Lorentz transformations – form a 10-parameter Lie group, the Poincaré group. In accordance with Section 4.2.2 we call the transformations $T_{II'}(v, \omega)$ also "generalised Lorentz transformations" since the numerical value of the parameter ω is still open.[15] As already mentioned in Section 4.2.2, with this last step we also arrive at the space – time continuum of special relativity, the Minkowski space M. The Minkowski space is a 4-dimensional manifold M that is equipped with a metric tensor $g_{\mu\nu}$. The signature of $g_{\mu\nu}$ is 2 and thus $g_{\mu\nu}$ is a Lorentz metric.[16] With this indefinite Lorentz metric on the manifold M, the non-zero vectors at a point p can be divided into three classes, into vectors that are called *timelike, null,* or *spacelike*. In the space T_p of tangent vectors of M at p, the null vectors constitute a double cone, the *null cone*, which separates the timelike from the spacelike vectors. (In order to avoid any terminological confusion, we do not use the term *light-cone* here).

In the coordinates (t, x_1, x_2, x_3) the *null cone* can be expressed by the equation

$$\omega^2 t^2 - x_1{}^2 - x_2{}^2 - x_3{}^2 = 0.$$

Hence the apex angle γ of the *null cone* (from the t-axis) is connected with the constant ω by the relation $\gamma = \text{arc tan}\ \omega$ or $\omega = \tan \gamma$ (Fig. 4.1). This connection shows that the constant ω, which could not be determined within the framework of our reconstruction of special relativity, is a genuine property of the Minkowskian space M. Since M is an empty space-time, the constant ω characterises an intrinsic feature of the Minkowski space, the *null cone* structure and that irrespective of any physical processes that might happen in this space-time.

This is the meaning of the constant ω on the most abstract level of the Minkowskian space-time M. Consequently, we should clearly distinguish between the constant ω,

---

[15] Mittelstaedt (2006).

[16] Hawking and Ellis (1973), p. 38.

which turned out to be a structural constant of the Minkowskian space-time, and other constants that characterise some physical processes. Within the present context, the propagation of light is of particular interest. According to the reconstruction of special relativity and relativistic mechanics, which we have carried out in Sections (2.2) and (2.3), respectively, we can make use of the components $(p^0, \boldsymbol{p})$ of the four-vector p^μ of the momentum for formulating the quantity

$$p^2 := (p^0)^2 - (|\boldsymbol{p}|)^2 = m_0{}^2\omega^2,$$

which is invariant against generalised Lorentz transformations $T_{II'}(v, \omega)$. From this invariance relation we obtain the formula[17]

$$p^0 = \sqrt{\mathrm{m_0}^2\omega^2 + |\boldsymbol{p}|^2}$$

which allows to express the three-velocity $v^i{}_P$ of a particle P as function of the rest mass m_0 and the three-momentum $\boldsymbol{p}$ by

$$v^i{}_P = p^i/\sqrt{m_0{}^2 + |\boldsymbol{p}|^2/\omega^2}.$$

Hence, for vanishing restmass ($m_0 = 0$) the momentum dependence disappears and we get

$$v^i{}_P = \omega p^i/|\boldsymbol{p}|,$$

which means that the particle (photon) will move along the null cone with velocity $|\boldsymbol{v}_P| = \omega$. We add that according to our present knowledge, the photon is even the only zero-mass particle. Moreover, since the velocity of light can be measured by well known methods to be $v_L = c$, the numerical value of the constant ω can now be determined as $\omega = c$. In order to avoid any terminological confusion, we clearly distinguish here the *null-cone* as an intrinsic characteristic of the Minkowski space from the *light-cone*, which governs the propagation of light in vacuum. For this reason, we did not use the term *light-cone* in the preceding sections.

Concluding this discussion we can argue, that the apex angle of the *null-cone* of the Minkowskian space is not determined by the velocity c of light, but that the velocity c of light in vacuum is determined by the apex angle $\gamma = \text{arc tan}\ \omega$ of the *null-cone* of the Minkowskian space M. In other words, we argue in favour of a strict dualism of space-time and matter within the framework of special relativity. From a historical point of view, the opposite position seems to be induced by the operational approach to special relativity by Einstein and other scientists in the beginning of the twentieth century. We mentioned this approach already in

[17] Cf. Mittelstaedt (1995), p. 114 and Sexl and Urbantke (1992), p. 68.

Section 4.2.1 (PRELIMINARY REMARKS) as a way to understand why the empirical value of the velocity of light could perhaps have some influence on the structure of space-time. However, our considerations have shown that this is obviously not the case.

4.2.5 *One More Fundamental Question*

Even if we could identify the constant ω as a characteristic number of the Minkowskian space-time, we could not present a sufficient reason for its empirical value $\omega = c = 3 \cdot 10^8$ m/s. Hence, the question arises whether this number is a "contingent" value without any sufficient reason. However, at this point we must first ask and clarify what "contingency" means in the context of physics. Presently, we know only one situation in physics in which we observe values that are completely fortuitous: The outcome of a quantum mechanical process can in general not be determined strictly but only by means of probabilities. This implies the following consideration. If c is a fundamental and universal constant that refers to all processes and to the entire world, and if c has a contingent value, then for an adequate understanding of the contingency of c we must consider the evolution of the universe and trace it back to the quantum mechanical process of the creation of the world in its beginning.[18] Since a quantum process of this kind will hardly create a single universe but rather a huge number of possible worlds, a so called "multiverse",[19] the fortuitousness of the fundamental constant c would correspond to a statistical distribution of possible values of c in the presumably large ensemble of simultaneously created worlds.

Hence, on the large scale between a small positive value $c > 0$ and the limit $c \to \infty$, which corresponds to Newton's classical physics, the value $c = 3 \cdot 10^8\, m/s$ determines the position of our actual Minkowskian space-time. Consequently, for any observer in our world, the constant c has a definite value that does not depend on any law within this world and not on the methods of observation. The constant c has a contingent value in our world, but in the large ensemble, the multiverse of simultaneously created worlds, there are merely probabilities for the many possible values of the so-called constant c.

4.3 Planck's Constant $\hbar$ in the Light of Quantum Logic

4.3.1 *Ontological Preliminaries*

A similar situation as in the case of the constant c can be found in the domain of quantum physics with respect to Planck's constant $\hbar$, which is widely considered as

[18] Vilenkin (1982) and Linde (1990).

[19] Ellis et al. (2003) and Carr (2007).

a characteristic of quantum mechanics. First, we mention that according to the results of Chapter 3, quantum mechanics can be obtained by a careful relaxation of some ontological hypotheses of classical mechanics. In addition, there are good reasons to assume that quantum mechanics is universally valid. Hence, we do not expect to find in quantum mechanics explicit indications of a borderline that is determined by $\hbar$ and that separates the validity domain of quantum mechanics from the validity domain of classical mechanics. The classical roots of quantum mechanics are completely eliminated in our way of reasoning and Bohr's correspondence principle has become void. Obviously, this situation is very similar to the corresponding investigations of the constant c in special relativity. However, from a more technical point of view there are important differences between the considerations about c and about $\hbar$. The starting point for a rational reconstruction of quantum mechanics is not a relaxed version of classical mechanics but a formal scientific language and logic that is based on a weak quantum ontology which is free from many hypothetical assumptions of a classical language. The corresponding formal logic, the "quantum logic" is the basis of a rational reconstruction of quantum mechanics in Hilbert space, which was discussed in Section 3.5.

In the present section we will investigate the important and at first sight surprising observation that in the well known systems of quantum logic Planck's constant $\hbar$ does not appear. What is the reason for this apparent deficiency of quantum logic? We recall that the main goal of quantum logic is the "bottom-up" reconstruction of Hilbert lattices, effect algebras, and of quantum mechanics in Hilbert space as treated in Section 3.5 – and all that without any reference to the actual historical development of the theory.[20] The starting point is a weak quantum ontology that describes the most general features of the quantum physical reality. As in Section 3.5, we consider here three types of ontologies of different strength: The classical ontology $O(C)$, the weaker quantum ontology $O(Q)$, and the unsharp quantum ontology $O(Q^U)$, which is partly stronger than $O(Q)$, since it allows for unsharp joint properties and partly weaker than $O(Q)$, since it does not require value definite properties. In any case, $O(Q^U)$ is weaker than the classical ontology $O(C)$. As to the terminology of weak and strong ontologies, we refer to the definition of a partial ordering relation between two ontologies in Section 1.1. Many details about these ontologies can be found in the literature.[21] The main result consists of the observation that the quantum ontologies $O(Q)$ and $O(Q^U)$ can be obtained by convenient relaxations of the classical ontology $O(C)$. The relaxations in question consist in the elimination of hypothetical assumptions contained in $O(C)$ that are neither justified by rational reasoning nor by experimental evidence. However, the present considerations will show that it is not sufficient to simply eliminate a certain hypothesis since it must be replaced by some weaker requirement.

[20] Dalla Chiara and Giuntini (2001).

[21] Mittelstaedt (2005).

Except for other assumptions of the classical ontology,[22] which are not relevant here, for the present investigations the classical ontology is characterised by the following requirements:

- There are individual and distinguishable objects S_i which possess elementary properties P_λ such that either P_λ or the counter property $\bar{P}_\lambda$ pertain to the system.
- Objects S_i are "thoroughgoing determined", i.e. an object possesses each elementary property either affirmative (P) or negative ($\bar{P}$). Hence, objects can be individualized by elementary properties and re-identified at later times if the property of impenetrability is presupposed.

There are important objections against this ontology $O(C)$, in particular against the second requirement. It is merely a hypothetical assumption that cannot be justified by rational arguments or by experimental evidence. In addition, classical ontology is not in accordance with quantum physics. A quantum system does not possess all elementary properties either affirmative or negative. Instead, only a subset of properties pertains to the system and can simultaneously be determined. These "objective" properties pertain to the object like in classical ontology. However, quantum objects cannot be individualised and re-identified by their objective properties, since there are not enough such properties. We will not use these material results here, but we learn from these considerations that classical ontology has too much structure compared with quantum physics. This observation offers the interesting possibility to formulate the ontology of quantum physics by relaxing some hypothetical assumptions of the classical ontology $O(C)$. We emphasise again that no new requirements will be added to the assumptions of $O(C)$. In this sense, our first attempt at a new quantum ontology $O(Q)$ reads:

- If an elementary property P pertains to an object as an objective property, then a test of this property by measurement will lead with certainty to the result P.
- Any elementary property P can be tested at a given object with the result that either P or the counter property $\bar{P}$ pertains to the system.
- Quantum objects are not thoroughgoing determined. They possess only a few elementary properties either affirmative or negative. Properties which pertain simultaneously to an object are called "objective" and "mutually commensurable".

The first two requirements are in complete accordance with $O(C)$ whereas the third one is a strong relaxation of the corresponding assumption of $O(C)$. However, as already mentioned in Section 3.5, the new quantum ontology $O(Q)$ is not yet in complete accordance with quantum physics for two reasons.

- The most general observables in quantum mechanics correspond to unsharp properties that allow for joint properties even for complementary observables.

[22] Cf. Section 1.3.

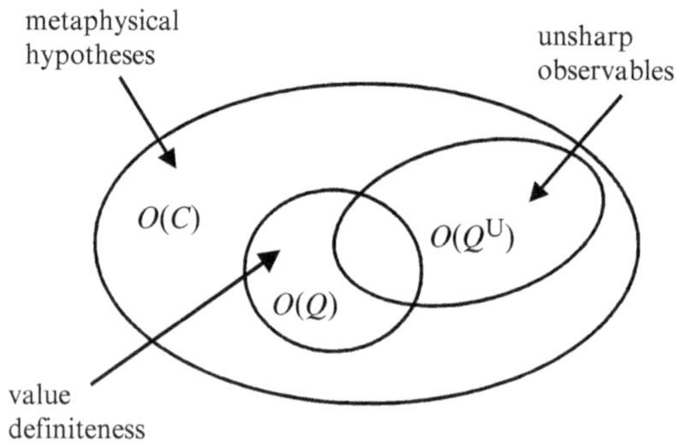

Fig. 4.2 Interrelations between the ontologies $O(C)$, $O(Q)$ and $O(Q^U)$

Hence, $O(Q)$ is too restrictive since it does not allow generally for joint properties.

- The requirement of value definiteness cannot be fulfilled for all properties since the pointer-objectification in the measurement-process cannot be achieved generally. Hence, $O(Q)$ is also not sufficiently restrictive.

These two objections against $O(Q)$ can both be taken into account if $O(Q)$ is replaced by a new quantum ontology $O(Q^U)$ of unsharp properties. It is a difficult to say how much unsharpness is needed for removing the two deficiencies of $O(Q)$ mentioned. We come back to this question at the end of Section 4.3.4.

Comparing the three ontologies in question, we find that on the one hand $O(Q^U)$ is partly stronger than $O(Q)$, since it allows for unsharp joint complementary properties that are not contained in $O(Q)$. On the other hand, $O(Q^U)$ is partly weaker than $O(Q)$, since value definiteness of properties is not required. However, $O(Q^U)$ is weaker than the classical ontology $O(C)$. These relations are illustrated again in Fig. 4.2.

4.3.2 *The Quantum Logic Approach*

The main goal of the quantum logic is the reconstruction of Hilbert lattices and of quantum mechanics in Hilbert space on the basis of the weak quantum ontologies mentioned.[23] Starting from the weak ontology $O(Q)$ we can construct a formal language S_Q of quantum physics whose syntax leads together with a convenient semantics of truth to the calculus L_Q of quantum logic. The Lindenbaum-Tarski algebra of L_Q turns out to be a complete, orthomodular lattice L_Q, which in addition is atomic and fulfils the covering law if the language is assumed to refer to a single system. We denote this lattice by L_Q^*. Using the Piron-McLaren Theorem[24] and the

[23] Dalla Chiara and Giuntini (2001).

[24] Mac Laren (1965), Mittelstaedt (2005) and Piron (1976).

angle-bisecting condition of Solèr,[25] we arrive at the three classical Hilbert spaces and in particular at the complex numbers Hilbert space $H(C)$ of quantum mechanics.

Compared with the classical ontology $O(C)$, a formal classical language S_C and the classical propositional logic L_C, there are important differences that come from the elimination of the hypothetical assumptions contained in $O(C)$. In particular, we have sacrificed here the assumption that objects are always "thoroughgoing determined". As a consequence of this reduction, propositions of the language S_Q loose their "unrestricted availability" and are in general only restrictedly available. For the calculus L_Q of quantum logic this relaxation implies the loss of the distributive law. We could go one step further and proceed to the ontology $O(Q^U)$ of unsharp properties by omitting the assumption that for each property P it is objectively decided whether P or its counter property $\bar{P}$ pertains to the system.

This relaxation implies that propositions are no longer value definite and that both the "excluded middle" and the "law of contradiction" are no longer formally true.[26] The theory which we obtain in this way by reducing ontological premises is an abstract Hilbert space quantum theory of sharp and unsharp properties. It is an empty theory, a formal framework of quantum mechanics, which is presumably universally valid. It is, however, not a priori valid in the strict sense, since the underlying ontologies $O(Q)$ and $O(Q^U)$ do still contain hypothetical premises that are not queried here. The abstract quantum theory which is reconstructed on the basis of the weak quantum ontology $O(Q^U)$ is, however, closer to the truth than the theory based on $O(Q)$ and in any case closer to the truth than the classical mechanics based on the classical ontology $O(C)$.

4.3.3 In Search of Planck's Constant

Within the quantum logic approach quantum mechanics in Hilbert space appears as an abstract and empty theory which is based on the weak quantum ontology and thus presumably universally valid. Hence, we expect first of all to discover somewhere in this theory Planck's constant $\hbar$, which is widely considered as a characteristic of quantum mechanics and as a number that indicates the border line between the quantum world and the classical world. However, within the quantum-logic approach there is no classical world[27] and hence no border line between the two worlds from which we could read off Planck's constant. Hence, there is no hope to find the constant $\hbar$ within the domain of abstract quantum theory in Hilbert space. In order to discover Planck's constant in the realm of quantum logic, we must extend the abstract and empty theory by incorporating real entities into the theory.

---

[25] Solèr (1995).

[26] Dalla Chiara (1995) and Foulis and Bennett (1994).

[27] Mittelstaedt (2005).

We will find that "objects" or "particles" can be comprehended if the abstract theory is extended by concepts that are usually considered as classical notions. Since intuitively, particles are objects that are somehow localised in space, we consider first the concepts of *localizability* and *homogeneity*.

4.3.3.1 Localizability

Let Δ be a domain of the physical space R. If $R = R^{(1)}$ is one dimensional, the domains Δ considered are Borel sets on the real line, i.e. $\Delta \in B(R)$. Let $L(H)$ be the set of bonded linear operators on a Hilbert space H. The mapping

$$E : B(R) \rightarrow L(H); \Delta \rightarrow E\{\Delta\}$$

is a projection valued measure (PV-measure), if $E\{\Delta\} = E\{\Delta\}^* = E\{\Delta\}^2$ for all $\Delta \in B(R)$, $E\{\mathrm{I}\} = 1$, and $E\{\cup \Delta_\mathrm{i}\} = \sum E\{\Delta_\mathrm{i}\}$. According to the spectral theorem, this PV-measure leads to a self-adjoint operator, the position operator Q. More generally, we could start with a non-empty set Ω, a σ-algebra F of subsets of Ω, and hence on a measurable space (Ω, F).A normalised positive operator valued measure (POV-measure) can then be defined by

$$E : F \rightarrow L(H) on\ (\Omega, F),$$

where $E\{X\} \geq 0, X \in F, E\{\Omega\} = I$ and $E\{\cup X_\mathrm{i}\} = \sum E\{X_\mathrm{i}\}$ for disjoint sequences $(X_\mathrm{i}) \in F$.

4.3.3.2 Homogeneity

Homogeneity and isotropy are features of the physical space which show that the physical space has no observable properties. In a one dimensional space this means that a translation by an amount α

$$g_\alpha : \Delta \rightarrow g_\alpha(\Delta) = \{\lambda : (\lambda - \alpha) \in \Delta\}$$

with $\Delta \in B(R)$, is a symmetry transformation. If the physical space is homogeneous then there exists a unitary operator U_α, depending on α, U_α: $E \rightarrow U_\alpha{}^{-1} E\, U_\alpha$ such that

$$E\{g_\alpha(\Delta)\} = U_\alpha{}^{-1} E\{\Delta\} U_\alpha$$

where $E\{\Delta\}$ is the projection operator mentioned above. We can choose the parameter α such that it is additive, i.e. $U_\alpha U_\beta = U_{\alpha+\beta}$. According to Stone's theorem and under this condition a self-adjoint operator P with $U_\alpha = \exp(i\alpha P)$, the displacement operator, is uniquely determined.

4.3.3.3 Canonical Commutation Relations

If we consider the position operator Q as generator of a one-parameter group with parameter β, we get $V_\beta = \exp(i\beta Q)$. Together with the corresponding expression $U_\alpha = \exp(i\alpha P)$ for the displacement operator P, we find the canonical commutation relations in the Weyl formulation

$$U_\alpha V_\beta = e^{i\alpha\beta} V_\beta U_\alpha$$

For a dense subset D of the entire Hilbert space the Weyl commutation relations imply that the operators P and Q satisfy the relation

$$[Q,P]f = iIf \quad \text{for all } f \in D.$$

4.3.3.4 Physical Objects

The notion of an object is intimately related with the equivalence of active and passive space-time transformations and with the covariance of observables under these transformations. Indeed, if we understand by an object an entity of the external reality that exists objectively and independent of the observing subject and his measurement devices, then it should not matter whether the object is (actively) transformed by a translation in space, say – or whether the apparatus and its coordinates are (passively) transformed in the opposite direction. If we combine this idea with the concepts of localization and homogeneity mentioned, we can characterize an "object" in the following way:
Let M be a topological space, the configuration space of the intended object and G a locally compact transformation group that acts transitively on M. Here, we think preferably of the Galilei group G and its one-parameter subgroups of space translations and velocity boosts. An element $g \in G$ induces a one-to-one and continuous mapping of M to itself

$$g : \Delta \to g(\Delta), M \to M,$$

with $\Delta \in B(M)$. A projection valued measure E: $\Delta \to E\{\Delta\}$ leads to a unitary representation of the group, $g \to U_g$ with

$$U_g : E \to U_g{}^{-1} E U_g.$$

We can now express the requirement of objectivity (of the intended object) by the covariance in Fig. 4.3.

For an element of the external reality it should not matter whether we transform first the domain Δ by mapping g of M to itself (passive transformation) and go in a

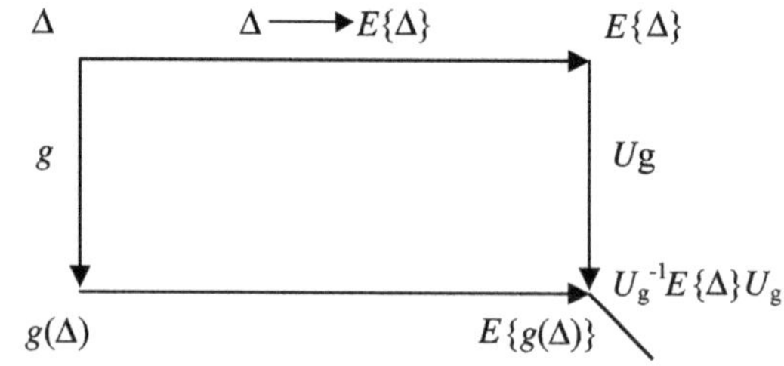

Fig. 4.3 Covariance diagram

second step (corresponding to a PV-measure) from $g(\Delta)$ to $E\{g(\Delta)\}$;– or whether we go first from Δ to the projection operator $E\{\Delta\}$ and in a second step (corresponding to a unitary representation U_g) from $E\{\Delta\}$ to $U_g^{-1}E\{\Delta\}U_g$ (active transformation).

This requirement, which expresses the equivalence of active and passive transformations, means that the covariance diagram in Fig. 4.3 commutes, i.e.

$$E\{g(\Delta)\} = U_g^{-1}E\{\Delta\}U_g$$

Observables E that fulfil this "covariance postulate" correspond to properties of he object in question that transforms covariantly under the transformation of the Galilei group. In other words, a quantum object is carrier of properties $E\{\Delta\}$ which transform under the Galilei transformations.[28]

4.3.3.5 Elementary Particles

On the basis of this general concept of a quantum object as carrier of properties $E\{\Delta\}$ of the orthomodular lattice L_Q, we can specify this concept by considering different classes. Different representations $g \to U_g$ of elements $g \in G$ of the 10-parameter Galilei group by automorphism U_g (on the lattice L_Q of projection operators) correspond to different kinds of objects. In particular, the elementary objects are given by irreducible representations $g \to U_g$. However, there are no irreducible unitary *true* representations $g \to U_g$ of the Galilei group but only projective ones that contain a real, yet undetermined parameter z.

It should be emphasised here, that this way of reasoning for the constitution of objects is not restricted to PV-measures i.e. to sharp observables, since it can easily be extended and generalised to unsharp observables in the sense of POV – measures. Unsharp properties are then given by "effects" in Hilbert space, the algebra of which will be denoted here by $E(H)$.[29] Hence, in the covariance diagram (Fig. 4.3) the PV – measures must be replaced by POV-measures and the lattice of

[28] Mittelstaedt (1995) and Piron (1976).

[29] Foulis and Bennett (1994) and Busch et al. (1995), p. 25.

projection operators by the effect algebra $E(H)$. In this case, objects are carriers of the most general observables, given by POV- measures.[30]

For further illustrating this result, we mention briefly a few technical steps.[31] From the position operator Q a self-adjoint operator Q' for the velocity can be obtained by formal differentiation with respect to the time t, i.e. by $Q' = i\,[H, Q]$, where H is the evolution operator which is not yet fully determined at this point. For motivating the next step, we consider again the classical situation. If we change the reference system K to a new system K' which moves with the constant velocity $v = v(K, K')$ relatively to K, the velocity Q' and the position operator Q will change according to the transformation.

If also in quantum physics this velocity boost transformation is considered as a symmetry transformation, then there exists a one-parameter unitary group G_v such that

$$Q' + v = G_v Q' G_v^{-1} \textit{ and } G_v G_{v'} = G_{v+v'}.$$

Since the system is elementary and G_v commutes with Q, we may write $G_v = \exp(iv\,f(Q))$ where f is a Borel function on the real line. For combining the velocity boosts with the displacements mentioned above we define a two-parameter family of unitary operators $T(\alpha, v)$ such that

$$Q + \alpha = T(\alpha, v) Q\, T^{-1}(\alpha, v)$$

$$Q' + v = T(\alpha, v) Q' T^{-1}(\alpha, v)$$

Hence, $T(\alpha, v)$ is a projective representation of the two-dimensional translation group whose arbitrary phase factor can be written in the form

$$T(\alpha_1, v_1)T(\alpha_2, v_2) = \exp\,\{i\, z/2(\alpha_1 v_2 - \alpha_2 v_1)\; T(\alpha_1 + \alpha_2, v_1 + v_2)$$

where $z \neq 0$ is an arbitrary real constant which distinguishes different inequivalent projective representations. We can re-identify the one-parameter subgroups U_α and G_v by the relations $U_\alpha = T(\alpha, 0)$ and $G_v^{-1} = T(0, v)$, and by means of the commutation relations we find $U_\alpha\, G_v^{-1} = \exp(i\, z\alpha v)\, G_v^{-1} U_\alpha$ and $G_v^{-1} = \exp(i\, z\alpha Q)$. From this relation we obtain in a few steps[32] $Q' = P/z$ and $P/z = i\,[H, Q]$ where H is the most general evolution operator which is compatible with the principle of Galilei invariance.

In order to identify the parameter z, we refer to some classical aspects of an elementary object. If the object considered is localizable, we will call it an elementary particle. The parameter z is often interpreted as the inertial mass m of the

[30] Busch et al. (1995), p. 52 and Mittelstaedt (1995).

[31] Jauch (1968).

[32] Jauch (1968).

particle in question. This is, however, not quite correct. A more detailed investigation that refers to the classical motion and to the dynamics of the particle shows, that $z = m/\hbar$, where m is the inertial mass and $\hbar$ is a universal constant. Indeed, if we describe the classical motion by the movement of a point in the configuration space (given by the real line), then we can identify the classical motion with the motion of the expectation value x of the position operator Q.

Let W be a state operator with $W \in T(H)_1{}^+$, where $T(H)_1{}^+$ is the set of *positive trace one operators*. Then we have $x = \mathrm{tr}\{WQ\}$, and for the velocity we find

$$dx/dt = tr\{WQ'\} = i \cdot \mathrm{tr}\{W[H, Q]\} = \mathrm{tr}\{WP\}/z.$$

Here we made use of the relations $Q' = P/z = i\,[H, Q]$ mentioned above. For the momentum we obtain

$$m\,dx/dt = \mathrm{tr}\{WP\}\,m/z = \hbar \cdot tr\{WP\}.$$

At this preliminary stage of the discussion the constant $\hbar$ can be identified as connecting the displacement operator P with the momentum operator p of the particle such that $p = \hbar P$ and can be determined experimentally (at least in principle) as

$$\hbar = m/z = 1.05 \times 10^{-27}\mathrm{erg\ s}.$$

Summarizing these arguments we find, that the decisive steps in our search for Planck's constant $\hbar$ made use explicitly of classical concepts:

1. The concepts of localizability and homogeneity of the physical space
2. The definition of objects making use of covariance diagrams that distinguish explicitly representations of the Galilei group by transformations of the physical space-time and by automorphism U_g on the algebra of projection operators.
3. The classical concept of movement of a point in the configuration space.

Hence, our first, still preliminary result is that for discovering the physical meaning of Planck's constant $\hbar$, in addition to the abstract quantum logic, classical concepts must be taken into account. However, this result would invalidate the idea of an autonomous quantum world without any recourse to a classical world.

4.3.4 The Meaning of $\hbar$ in the Quantum World

Within the framework of the quantum theory in Hilbert space, we can derive a relation that is of particular importance for the constant $\hbar$. We think of the uncertainty relation, in particular if it is formulated in terms of unsharp observables, i.e. of POV-measures,[33] On the basis of the requirement of Galilei covariance for position q and momentum p the uncertainty relation

$$\delta q \cdot \delta p \geq \hbar/2$$

can be derived, where the meaning of the expressions δq and δp differs in various interpretations of quantum mechanics. Here we are interested in the "Heisenberg interpretation" of the uncertainty relations, i.e. an individualistic interpretation of these relations in terms of unsharp observables q and p.[34] The number $\hbar$ may then be considered as the *smallest possible degree of inaccuracy of jointly measured observables q and p that are probabilistically complementary*. Obviously, this meaning of $\hbar$ can be expressed exclusively in terms of quantum physics and without any recourse to the classical world.

As to quantum-logic, the meaning of $\hbar$ must be expressed in terms of the abstract language S_Q and of the formal logic L_Q. Compared with the language S_C of classical physics, the main restriction of quantum language is the restricted availability of propositions in a formal proof process. Proof processes are formulated either by a derivation within the framework of a calculus or by a dialog according to the rules of the material or formal dialog game. If after a material proof of a proposition A another proposition B was successfully shown to be true, the previously proved proposition A is no longer available except when proposition A and B are commensurable. For unsharp propositions, these strict alternatives can be considerably relaxed, since even probabilistically complementary, unsharp propositions are not strictly incommensurable. Consequently, the degree to which the proposition A is still available after a proof of B depends on the degree of commensurability of A and B. Hence, in the spirit of the uncertainty relation for individual unsharp propositions, the constant $\hbar$ can be identified here as a measure for *the smallest possible unavailability of the unsharp complementary propositions A and B*. In other words, Planck's constant determines the smallest possible unavailability, and in this sense $\hbar$ is a universal constant in the realm of quantum-logic.

We can go one step further to the ontology $O(Q^U)$ of unsharp properties. Compared with the classical ontology $O(C)$, the main restriction of the quantum ontology $O(Q)$ is, that objects are not thoroughgoingly determined. This restriction is, however, too strong since for the most general observables we must allow for unsharp joint properties even for probabilistically complementary observables. This argument is taken account of in the ontology $O(Q^U)$ of unsharp properties. Objects of $O(Q^U)$ are "partially determined" in the sense that all possible predicates pertain at least unsharply to the system. This is, however, only a qualitative characterisation of the ontology $O(Q^U)$. Quantitatively, and in the spirit of the uncertainty relation we can say that the *minimal degree of unsharpness of probabilistically complementary*

[33] Busch (1985) and Busch et al. (1995, 2007).

[34] Busch et al. (2007).

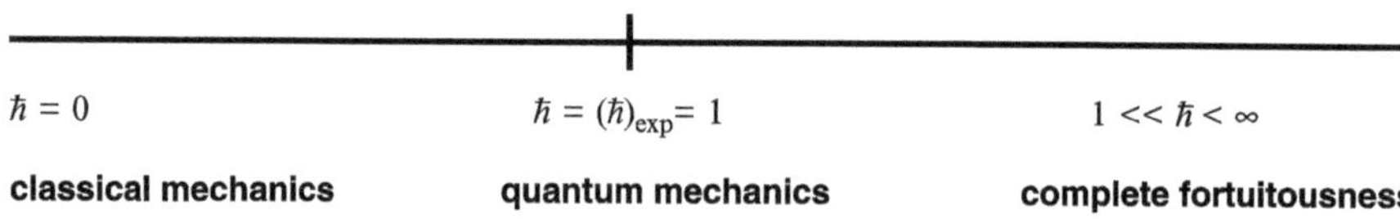

Fig. 4.4 Planck's constant shows the position of quantum mechanics

properties which pertain jointly to a system is given by $\hbar$. This is the meaning of Planck's constant on the level of quantum ontology (Fig. 4.4).

Hence, we find that Planck's constant $\hbar$ is also an intrinsic characteristic of quantum ontology and thus of the quantum world at all, that describes the largest possible degree of joint determination of unsharp complementary properties.[35] In contrast to classical systems, quantum systems are only partially determined, where the largest possible degree of partial determination is measured by $\hbar$. Since in this way we can identify $\hbar$ as an intrinsic feature of quantum physics and express this feature exclusively in terms of quantum physics, we could dismiss the long detour on our way to Planck's constant making use of several classical concepts. In other words, we can throw away "Wittgenstein's ladder" whose steps consist in the present case of various classical concepts. On the basis of these results, we can now try to answer the question posed in the introduction: What is the reason why in the operational approach to quantum logic the constant $\hbar$ does not appear in the first instance? For a bottom-up reconstruction of quantum mechanics the reduced quantum ontologies $O(Q)$ and $O(Q^U)$ are too general and not sufficiently specific for a complete reconstruction of quantum logic and quantum mechanics. These theoretical structures can be obtained only up to an unknown real parameter, whose numerical value must be determined empirically and turns out to be $\hbar$. Hence, on the long scale between classical physics and complete fortuitousness, Planck's constant determines the actual position of quantum logic and quantum physics.

4.4 The Problem of the Gravitational Constant κ

In Sections 4.2 and 4.3 of the present chapter we could show that the two constants of nature c and $\hbar$ don't describe border lines between the Newtonian physics and special relativity and quantum mechanics, respectively, – but instead intrinsic properties of the Minkowskian space-time and of the quantum world. In analogy, one could guess that the relativistic gravitational constant κ is not only a borderline between Newton's theory of gravitation and the domain of very strong gravitational fields as they are treated in general relativity, but a characteristic intrinsic feature of the pseudo-Riemannian space-time which is generated by the distribution of matter and by appropriate boundary conditions.

[35] Cf. Mittelstaedt (2008)

In general relativity, many properties of the Minkowskian space-time have only a local meaning, as the local and momentarily systems of inertia on the world-line of an observer and the local null-cones at any space-time point of such a time-like world-line. Graphical illustrations of these features are well known from the literature in general relativity. (Cf. for instance Hawking and Ellis (1973), p. 152 for the Schwarzschild solution). Hence, we expect that the apex angles of the null-cones on the world-line of a geodesic observer depend not only on the constant c as in special relativity, but also on the gravitational constant, which determines the strength of the local gravitational field, i.e. the curvature of the space-time at the space-time point in question. In this sense, the gravitational constant κ describes an important intrinsic property of the pseudo-Riemannian space-time – and that without any recourse to special relativity and to the Newtonian space-time.

These considerations show that also the gravitational constant κ – in a similar way as the constants c and $\hbar$ – may be considered as a characteristic of the space-time structure of general relativity. This impression is completely correct if we forget about the origin and the development of general relativity and take this theory simply for granted. The same meaning of the constant κ is confirmed also by the field-theoretical flat-space-time approach to general relativity, which considers κ as the coupling constant between the gravitational field – given by a symmetric tensor field $\psi_{\mu\nu}$ and the field generating energy-momentum (matter) tensor $T_{\mu\nu}$. However, against this interpretation of the meaning of the constant κ, in important argument can be put forward. General relativity can be reconstructed by relaxing hypothetical assumptions of Newton's classical physics only partially. As shown in Section 2.6 we can obtain in this way merely the pseudo-Riemannian structure of space-time but not Einstein's field equations, which contain the gravitational constant κ and import it into the full theory of general relativity.

This means, that the general way of reasoning in the present treatise cannot be applied to the constant κ. In the case of the constants c and $\hbar$ we could argue, that the theories of special relativity and of quantum mechanics can be reconstructed merely by relaxing several hypothesis of Newton's classical physics and that without any new empirical results. In this approach, the constants c and $\hbar$ appear as characteristics of the two resulting theories, which show intrinsic properties of the relativistic space-time and of the quantum world, respectively. Since this way of reasoning is not applicable to general relativity, the constant κ has at first merely the meaning of an experimentally well established coupling constant.

In Section (2.6) we discussed the attempt to reconstruct general relativity. In this context we mentioned Wheeler's six hypothetical approaches to Einstein's field equations. If one approach of this six-fold way to general relativity were successful, then we could say something about the meaning of the constant κ. Indeed, on the basis of the full theory of general relativity, i.e. the pseudo-Riemannian space-time and Einstein's field equations, the mere coupling constant κ obtains a much deeper meaning. This meaning can be expressed in different ways by the influence of the energy-mass density on the curvature of the Riemannian space-time. In the literature[36] we find an extensive discussion of the various ways to this goal. Perhaps the most fundamental and at the same time the most intuitive way is the approach (no. 6) by Sakharov,[37] which interprets the constant κ as a measure of the "metric elasticity of space", as a characteristic constant that describes the resistance of space to deformations. The conceptual framework of this view is well known from classical mechanics of elastic continua. The resistance of a homogeneous isotropic solid to deformations is usually described by two elastic constants, Young's modulus and Poisson's ratio. Here, the resistance of space to deformations is described by only *one* constant, the constant of gravity.

4.5 Three Constants of Nature – Concluding Remarks

In the preceding Sections 4.2, 4.3, and 4.4 of the present chapter, we discussed the physical meaning of the three fundamental constants c, κ, and $\hbar$ and that in the light of the rational reconstructions of modern physics presented in this treatise. The three constants in question refer to the three major theories of modern physics, to *special relativity, general relativity,* and *quantum mechanics.* Accordingly, the usual interpretation of the constants mentioned is to indicate the borderlines between the domains of the three theories of modern physics and the domain of classical physics. As mentioned already in Section 4.1, this well known interpretation is, however, not tenable. The reason is that in the rational reconstructions of modern physics we eliminated the classical structures and all traces of the classical world. Hence, it is very hard to believe that the well established theories of modern physics contain constants of nature that indicate the borderlines to a fictitious classical world and to the pretended classical roots of modern physics.

For these reasons we made a fresh start to understand the meaning of the three constants of nature and from now on exclusively within the frame work of the theories of modern physics. In this way, it turned out, that the three constants c, κ, and $\hbar$ characterise essential intrinsic features of the domains of validity and application of the three theories in question. Of course, we obtained these results without any explicit reference to classical mechanics and classical space-time.

(a) *Special relativity* is concerned with physics in the 4-dimenional Minkowskian space-time, which is equipped with a Lorentz metric of signature 2. The constant c characterises an intrinsic feature of this space-time, the apex angle γ of the null cone, which is related to the constant c by $\gamma = \text{arctg}\ c$.
(b) *General relativity* is concerned with a matter induced Pseudo-Riemannian space-time of signature 2. The gravitational constant κ is a characteristic feature of this space-time and describes the "metric elasticity of space".

[36] Misner et al. (1973), pp. 417–428, in particular p. 426.
[37] Sakharov (1967).

(c) *Quantum mechanics* does not allow to attribute two complementary properties jointly to an object system, except the two properties are sufficiently unsharp. The constant $\hbar$ is a characteristic of the quantum world and describes the *largest possible degree of joint determination of two unsharp probabilistically complementary properties.*

Summarising these results we find that the constants c, κ, and $\hbar$ describe intrinsic features of the domains of *special relativity*, *general relativity* and *quantum mechanics*, respectively.

Chapter 5
Interpretations of Modern Physics

5.1 Introductory Remarks

Since the advent of modern physics in 1905, when Einstein's theory of special relativity appeared, we observe a rapidly increasing activity to "interpret" this new and for the present somewhat strange theory of modern physics. However, it should be emphasised, that special relativity was only the first in a sequence of new theories, that allegedly required an "interpretation". It was followed by general relativity, which from a mathematical point of view is much more ambitious and thus even less comprehensible than special relativity. Accordingly, interpretations of general relativity are concerned with mathematical subtleties as well as with purely conceptual problems. The third theory in the sequence in question is quantum mechanics. With general relativity it shares the great mathematical complexity and intricacies, with special relativity the new conceptual situation, in particular the difficult interrelations between classical physics and the new theory. Hence, it should not be very surprising that the majority of interpretations of modern physics are concerned with quantum mechanics.

In contrast to the obvious activity in interpreting the three theories of modern physics, it is at first sight somewhat astonishing that comparable interpretations of classical physics, in particular of classical mechanics, are almost unknown. Only at the end of the nineteenth century we find at least three remarkable exceptions: First the critical investigations of Ernst Mach about the concepts of space and time in Newton's mechanics,[1] second the investigations of Henri Poincaré[2] about the conventionality of distant simultaneity and the various conventions underlying the measure of time in mechanics; and third, the investigations of H. von Helmholtz[3] about the empirical facts underlying the geometry of the three dimensional physical space. – The common goal of these and comparable investigations

[1] Mach (1901).

[2] Poincaré (1898).

[3] von Helmholtz (1868).

P. Mittelstaedt, *Rational Reconstructions of Modern Physics*,
Fundamental Theories of Physics 174, DOI 10.1007/978-94-007-5593-2_5,

was a careful distinction between experimental facts, conventions and hypothetical assumptions in classical mechanics. In Newton's *Principia* these components are not adequately distinguished. However, it will become obvious in the following sections that this kind of interpretation of classical mechanics is rather innocuous compared with the activities that we call "interpretations of modern physics".

The different ways of treating classical physics and modern physics raise the question, as to where the difference comes from. Not only in the beginning of the twentieth century, but also today the usual answer given to this question by physicists and philosophers of science reads: In contrast to classical physics, which is intuitive and comprehensible, modern physics is unintuitive and difficult to understand, and hence it requires an additional interpretation. Usually, "interpretation" is understood thereby as a way to establish some relations between the formal expressions of the theory and elements of physical reality. – However, in contrast to this wide-spread conviction we refer to the opposite way of reasoning that we formulated in Chapters 1, 2 and 3. If, according to the considerations in these chapters the theories of modern physics are at bottom intuitive and comprehensible, then we must ask again, why these theories require something like an "interpretation". We expect that a detailed investigation of this problem will show that for the theories of modern physics there is actually no need for additional interpretations. In order to confirm the arguments of these preliminary considerations in the light of Chapters 1, 2 and 3, we will investigate here the two most interesting interpretations that were applied to theories of modern physics. Thereby, we distinguish two types of interpretations, depending on whether the theoretical terms are interpreted (a) by entities inside the realm of the theory, and (b) by entities outside the realm of the theory. Accordingly, we distinguish interpretations with (a) internal semantics and (b) external semantics.

Since classical physics, and in particular classical mechanics, is usually considered as the physical theory that correctly governs our ordinary experience, also Newton's mechanics and his space-time theory are considered as intuitive and understandable without difficulties. However, this wide–spread conviction is not self-evident, but a risky hypothesis. In order to elucidate this sceptical estimation more explicitly, we recall the six hypothetical assumptions that are presupposed in classical mechanics and in the classical theory of space-time. These assumptions are formulated partly in Newton's *Principia* and in writings about metaphysics and theology of the seventeenth and eighteenth century. In Section 1.3 we called these assumptions the ontological hypotheses of classical mechanics. The six hypotheses $O(C)^1 \ldots O(C)^6$ of the classical ontology $O(C)$ read

$O(C)^1$ There exists an absolute time. It establishes a universal temporal order of two or more events, it provides a universal measure of time and it explains the concept of simultaneity of two spatially separated events.

$O(C)^2$ There exists an absolute space. It explains the concepts of absolute motion and absolute rest. Euclidean geometry applies to this absolute space.

$O(C)^3$ There are individual and distinguishable objects. These objects cannot only be named and identified at a certain instant of time, but also re-identified at any later time. Generally, they also possess a temporal identity.

$O(C)^4$ These objects possess elementary properties P_λ in the following sense: An elementary property P_λ refers to an object system such that either P_λ or the counter property $\overline{P}_\lambda$ pertains to the system. Furthermore, objects are subject to the law of thoroughgoing determination, according to which "if all predicates are taken together with their contradictory opposites then one of each pair of contradictory opposites must belong to it".[4]

$O(C)^5$ For objects of the external objective reality, the causality law holds without any restriction. There is an unbroken causality.

$O(C)^6$ For objects of the external objective reality the law of conservation of substance holds without any restriction.

Obviously, these hypotheses can be justified neither by rational reasoning nor by experimental evidence. They are mere hypothetical assumptions that are not strictly realised in our ordinary everyday experience. In addition, they are in conflict with the results of Modern Physics.

5.2 The Interpretation of Special Relativity

Special Relativity was not only the first of the theories of Modern Physics that allegedly required an additional interpretation, but from a formal point of view it is also a very simple theory. For these reasons, we begin our considerations with this theory. As already mentioned in Chapter 2, in particular in Section 2.5, the theory of Special Relativity confronted the scientific community with several surprising and astonishing results, which called for an additional and helpful interpretation of these results. According to the theory, there is

- no absolute simultaneity of spatially separated events
- a time dilatation of moving clocks
- a length contraction of moving bodies
- a maximal velocity of moving observers

These results mentioned are *surprising* and even *puzzling*, simply since they contradict the ontological hypotheses $O(C)$ of classical mechanics, in particular the hypothesis $O(C)^1$ of the existence of an absolute and universal time. Here we ask how these contradictions between the strange implications of Einstein's theory and the ontology $O(C)$ can be *removed, moderated,* or *explained* in some way. In principle, there are two ways:

(A) We could apply a convenient "interpretation" to Special Relativity that "explains" the results of this theory by an appropriate semantics.

[4] Kant (1998), B599.

(B) We could take into account that the classical ontology $O(C)$ is loaded with hypotheses that can neither be justified by rational reasoning nor by experimental evidence.

We begin with the option (A) and consider first the historically most important interpretation.

5.2.1 The Traditional Interpretation by Means of Light Signals

Since all the surprising results are assumed to be outcomes of measurements, Einstein made use of an important methodological requirement: the apparatuses for the measurements of time intervals and spatial distances must be truly existing physical objects, such as clocks, measuring rods, etc. Of course, this postulate expresses clearly the difference to Newton's position in his *Principia*. We repeat the relevant passage quoted already in Section 1.1:

> Absolute, true and mathematical time, of itself, and from its own nature flows equably without regard to anything external, and by another name is called duration.[5]

In many interpretations of special relativity, the requirement mentioned was used and considered to be very helpful. It leaves, however, the question open whether the really existing measuring apparatuses are entities within the realm of the theory considered, or whether they are outside the domain of reality, that is described by the theory. – We will come back to this problem in the following considerations.

Already in the paper of 1905, Einstein defined the concept of distant simultaneity operationally by means of light signals. In a later publication,[6] which he considered to be intelligible to a wider audience, he made often use of light signals for explaining several thought experiments, e.g. the famous train experiment. This means that light signals were considered as instruments for measuring spatial and temporal distances. As already mentioned in Section 4.2.1, this way of understanding and interpreting special relativity was extended by the construction of a *light clock* many years later[7] and finally by an approach making use of particles and light rays.[8] Einstein had based the theory of special relativity on two postulates, the *principle of relativity* and the *principle of the constancy of the velocity of light*. According to the light postulate, the propagation of light is independent of the state of motion of the source of light.[9] This light principle proved to be very important for the "explanation" of many relativistic effects. It is, however, at first not a theorem of

---

[5] Newton (1934), p. 6.

[6] Einstein (1917).

[7] Marzke and Wheeler (1964).

[8] Ehlers et al. (1972).

[9] Einstein writes: "Das Licht hat im Vakuum stets eine bestimmte Ausbreitungsgeschwindigkeit, unabhängig vom Bewegungszustand der Lichtquelle". Cf. Stachel (2002), p. 107 and note 35.

special relativity. It must be justified either by another theory,[10] by the assumption that light is composed of photons, i.e. of zero-mass particles[11] or by experimental evidence.[12]

In the same way as Einstein, many other physicists used light signals for explaining and interpreting the various surprising effects of special relativity. As an example, we mention here the famous textbook by Max Born.[13] According to the above-mentioned requirement, light signals are indeed truly existing measurement instruments that can be used for the empirical determination of spatial and temporal distances and thus for an empirical confirmation of the entire theory. However, light rays, which are considered here as measuring apparatuses, are at first not subject to the theory of special relativity. Indeed, light as an electromagnetic wave phenomenon must be treated in Maxwell's theory, or even in quantum electrodynamics, if the photon structure of light is taken into account. In any case, the interpretation of special relativity in terms of light rays and light signals makes use of light-rays semantics in the sense defined above.

Hence, in this interpretation, the theory describes the physical reality as it appears, if we investigate it by means of light signals.[14]

It is obvious that this way of grasping the reality is not an approach in the sense of realism.

At this point the question arises as to why we actually need a separate interpretation in addition to the formal theory of special relativity. There are two reasons. At first, we cannot leave the theory as it is and relate its propositions directly to the physical reality since these propositions are partly in contradiction with our everyday experience, which we usually assume to be represented correctly by classical physics. For this reason, the contradiction in question comes from the incompatibility of special relativity and classical mechanics. Hence, the purpose of the present interpretation is first of all to remove this contradiction such that theoretical statements – like the time dilatation of moving clocks – are no longer in contradiction with our ordinary experience. This goal can be achieved in the present case by relating the propositions of the theory not directly to the physical reality, but to the image that we get grasping the reality by means of light rays. In this way, the results of special relativity lose much of their confusing character.

This last result leads directly to the second reason for using the interpretation considered. It was of particular importance in the first two decades after the

[10] E.g. by Maxwell's theoretical treatment of light as an electromagnetic wave phenomenon.

[11] Mittelstaedt (1976/1989), 3. ed. On pp. 124–126 it is shown, that only mass zero particles fulfil the light principle and vice versa.

[12] The first attempt of an empirical justification is de Sitter's analysis (1913) of double stars. Direct laboratory evidence was not known before the 1960s.

[13] Born (1920).

[14] Cf. Mittelstaedt (2011).

discovery of special relativity. Since during this period Einstein's theory had the bad reputation of being abstract and difficult to understand, it was subject to a clear rejection and even to political persecution.[15] The interpretation considered relates the propositions of the theory not directly to the physical reality but to the reality as it appears if it is investigated by light rays. Within the framework of this indirect "light rays interpretation", special relativity improves its plausibility, it loses much of its irritating character and there are no longer reasons for public agitation.

5.2.2 *Relaxation of the Classical Ontology O(C) – The Relativistic Ontology*

All these details are more or less known. There is, however, a completely different way to understand special relativity, that is based on the present investigations in Chapters 1 and 2. Instead of removing the contradiction between special relativity and classical mechanics by an indirect interpretation with a light rays semantics, we can proceed as follows: In a first step, we have to make clear that classical physics and the classical ontology $O(C)$ are neither intuitive nor plausible. Instead they are loaded with several hypotheses – in particular with the hypothesis $O(C)^1$ of the existence of an absolute time – which can neither be justified by rational reasoning nor by experimental evidence. In a second step, we have to abandon the hypothesis $O(C)^1$ of the existence of an absolute time and to relax the hypothesis $O(C)^2$ which states the existence of an absolute space. In this way, we obtain a less hypothetical "relativistic ontology" $O(SR)$, which follows from special relativity and is thus *necessary* for the theory *SR*. This ontology $O(SR)$ reads:

$O(SR)^2$ Euclidean geometry applies to the relative space.
$O(SR)^3$ There are individual and distinguishable objects.
$O(SR)^4$ These objects are thoroughgoingly determined.
$O(SR)^5$ The causality law holds for objects of the external material reality.
$O(SR)^6$ The law of conservation of substance holds for objects of the external material reality.

Obviously, this relaxed relativistic ontology $O(SR)$ does no longer contradict the surprising effects of special relativity. Indeed, only the eliminated hypothesis $O(C)^1$ would lead to a contradiction with the surprising relativistic effects. Without the hypothesis $O(C)^1$ and with the relaxed hypothesis $O(SR)^2$ we do not obtain any contradiction between special relativity and the ontology $O(SR)$. Here, we have removed the contradiction between special relativity and classical mechanics by abandoning the hypothesis $O(C)^1$ and by relaxing the hypothesis $O(C)^2$ of classical ontology. This shows that it is not special relativity that was responsible for the contradiction in question, but the hypothetical assumptions underlying classical mechanics.

[15] Könneker (2001).

Furthermore, we note that on the basis of the relaxed ontology $O(SR)$ the theory of special relativity can be reconstructed completely.[16] Hence, the ontology $O(SR)$ is also *sufficient* for the theory *SR*. With respect to our main topic, the interpretation of special relativity, this means that the outcomes of special relativity are results that can be understood literally. There is no need to "interpret" the propositions of special relativity by means of measuring apparatuses, i.e. in the present case, in terms of light rays and light signals. On account of the rational reconstruction in question, special relativity allows for a direct realistic interpretation. The public agitation about Einstein's theory in the first decades of the twentieth century was based on a fundamental misunderstanding of classical physics and classical ontology.

5.3 The Interpretation of General Relativity

As already mentioned in Section 2.6.2.1 a rational reconstruction of general relativity can be carried through merely with respect to the geometry of space-time in general relativity but not with respect to Einstein's field equations. Hence, if we are talking about general relativity, we have in mind only the space-time structure in finitely extended domains and also the global large scale structure of space-time. However, even this space-time geometry of general relativity requires an additional interpretation.

As already shown in Section 2.6.1 the space-time of general relativity turns out to be a 4-dimensional pseudo-Riemannian space-time with signature 2. This result implies a number of possible effects and phenomena of the large-scale structure of space-time that are not in accordance with our intuitive representations of space and time. Similar to the case of special relativity we are again confronted here with strange and puzzling results, which call for an additional interpretation. According to these considerations we mention here the following examples for global strange and puzzling results:

- The world could be a static universe
- the world could be an expanding universe
- the world could be a rotating universe
- the world could allow for closed time-like curves and thus for time travel
- the world could have an infinite age.

These possible results contradict our intuitive representations of space and time – and thus the ontological hypotheses $O(C)$ of classical mechanics, which are often assumed to govern our everyday experience. Hence we ask again how these contradictions between the puzzling results of general relativity and the ontology

[16] Mittelstaedt (1995, 2006).

O(*C*) can be removed or relaxed in some sense. As in the case of special relativity, there are in principle again two ways:

(A) We could apply a convenient "interpretation" to general relativity that "explains" the results of this theory by an appropriate semantics.
(B) As in the case of special relativity, we could take into account that the classical ontology *O*(*C*) is loaded with hypotheses that cannot be justified by rational reasoning or by experimental evidence.

We begin again with option (A):

5.3.1 The Traditional Interpretation by Means of Observations and Apparatuses

With respect to the results of general relativity, in particular in the global domain, an application of option (A) is possible but not fully satisfying. Even the most simple global phenomena as the isotropic expansion of the universe cannot directly be observed but only by means of high-tech apparatuses as optical telescopes, radio telescopes and other receivers for the cosmic microwave radiation. This means that Einstein's important requirement – that all empirical evidence must be obtained by truly existing apparatuses – is fulfilled *per se*. Hence, we arrive at a result that sounds similar to the corresponding result in Special Relativity:

> In the interpretation applied here the theory of general relativity describes the large-scale structure of space-time as it appears, if we investigate it by means of all kinds of telescopes and connected apparatuses.

Again it is obvious that this way of grasping the global reality of the universe is not an approach in the sense of realism. However, irrespective of the low epistemological value of this option, the contradictions between cosmological results of general relativity and our intuition of space and time disappear in this instrumentalistic interpretation. Hence, not only the present results but also the rapidly changing information about the global structure of our universe should not be considered any longer as reasons for public agitation. – Since a direct and realistic understanding of the global results of general relativity cannot be achieved by option (A), we will proceed now to the other option (B):

5.3.2 Relaxation of the Classical Ontology O(C) – The Ontology of General Relativity

Apart from the instrumentalistic interpretation of option (A) there is also here a completely different way to understand general relativity that is based in particular on the investigations of Section 2.6. Instead of removing the contradiction between possible geometric results of general relativity and our so – called intuitive understanding of space and time by an instrumentalistic interpretation, we can also proceed as follows: In a first step we have to make clear that classical physics and the classical ontology $O(C)$ are neither intuitive nor plausible. Instead they are loaded with numerous hypotheses – in particular with the hypothesis $O(C)^1$ of the existence of an absolute time – which can neither be justified by rational reasoning nor by experimental evidence. The hypothesis $O(C)^2$ consists of two parts, the assertion that there is an absolute space and the claim that Euclidean geometry applies to this space. We abandoned the assertion of the existence of an absolute space already in the ontology of special relativity but not the claim that Euclidean geometry applies at least to the remaining relative space. However, according to the reconstruction of the space-time of general relativity in Section 2.6.2.1 this claim can no longer be preserved. It was based according to Helmholtz' theorems on the assumption of a free mobility of finitely extended measuring rods. Hence, not only the hypothesis $O(C)^1$ but also the hypothesis $O(C)^2$ must be completely abandoned and thus we obtain the ontology $O(GR)$ of general relativity:

$O(GR)^3$ There are individual and distinguishable objects.
$O(GR)^4$ These objects are thoroughgoingly determined.
$O(GR)^5$ The causality law holds for objects of the external material reality.
$O(GR)^6$ The law of conservation of substance holds for objects of the external material reality.

This relaxed ontology of general relativity does no longer contradict the possible results of the large-scale space-time structure. Indeed, the four requirements of the weak ontology $O(GR)$ follow from the theory GR and only the eliminated hypotheses $O(C)^1$ and $O(C)^2$ could lead to contradictions with the possible results mentioned. Hence, with respect to the geometrical component the ontology $O(GR)$ is *necessary* for the theory GR. This shows that for the contradiction in question it is not general relativity that should be considered responsible but the hypothetical assumptions underlying classical mechanics.

Furthermore, we note that the relaxed ontology $O(GR)$ allows for a complete reconstruction of the "strange" space-time structure of general relativity and is thus also *sufficient* for this domain of the theory GR. With respect to our main topic, the interpretation of the space-time structure of general relativity, this means that the outcomes of general relativity are results that can be understood literally. There is no need to "interpret" the geometrical propositions of general relativity by means of measuring apparatuses, i.e. in the present case by means of all kinds of telescopes.

5.4 The Interpretation of Quantum Mechanics

Our third example of an interpretation of a theory of modern physics refers to quantum mechanics. Similar to the case of special and general relativity, quantum mechanics leads to results that are surprising and astonishing and not in accordance with classical mechanics and with the hypotheses of the classical ontology *O*(*C*). These strange and astonishing results of quantum mechanics reads:

1. A quantum mechanical object does not possess all elementary properties either positive or negative.
2. Only a subset of all properties pertains to the system and can simultaneously be determined.
3. These "objective" properties pertain to the system as in the classical ontology *O*(*C*).
4. For quantum objects no strict causality law can be established.
5. Quantum objects cannot be individualised and re-identified.

These results are considered to be "strange", "astonishing", and "surprising" since they contradict several hypotheses of the classical ontology *O*(*C*). Hence, the important question arises again as to how these contradictions can be removed. As in the case of Special and General Relativity, there are in principle two ways.

(A) We could apply a convenient "interpretation" to quantum mechanics that "explains" the results of this theory by an appropriate semantics.
(B) As in the case of special relativity, we could take into account that the classical ontology *O*(*C*) is loaded with hypotheses that can neither be justified by rational reasoning nor by experimental evidence.

We begin with the option (A) and consider first the two historically most important "interpretations" of quantum mechanics.

5.4.1 The Copenhagen Interpretation of Quantum Mechanics

Our first example is the so called "Copenhagen interpretation" of quantum mechanics. Usually, it is considered as Niels Bohr's interpretation of this theory. However, according to a very interesting historical investigation by Don Howard,[17] this first interpretation of quantum mechanics was called "Copenhagen interpretation" by Heisenberg not until 1935, i.e. 10 years after the formulation of quantum mechanics. We will not follow here this historical way of reasoning but instead consider at first Bohr's early complementarity interpretation. This interpretation is best characterised by the keywords "complementarity" and "correspondence" and by the recourse to classical apparatuses, classical concepts, and ordinary language. It was presented first by Bohr in his Como lecture of 1927, which was published in 1928.[18]

In the Copenhagen interpretation, Bohr made use again of the methodological requirement that was used already by Einstein in his investigations of special and general relativity: measuring instruments that are used for the interpretation

[17] Howard (2004).

[18] Bohr (1928).

of theoretical expressions must be truly existing physical objects. In this sense, Bohr always assumed that the apparatuses for measuring observable quantities like position, momentum, energy etc. can actually be constructed in a laboratory. By means of this methodological assumption, Bohr could explain one of the most surprising features of the new theory, which he called "complementarity". In quantum mechanics, two observables A and B that are canonically conjugate in the sense of classical mechanics, cannot be measured simultaneously. The most prominent example of this non-classical behaviour is the complementarity of the position q and the momentum p. Bohr explained the complementarity of the observables p and q in the following way: The measuring apparatuses $M(p)$ and $M(q)$ that could be used for measuring p and q, respectively, are mutually exclusive. In other words, there is no real instrument $M(p, q)$ that could be used for a joint measurement of p and q.

Hence, in the Copenhagen interpretation quantum mechanics assumes a semantics in the sense of instrumentalism and describes the microscopic quantum world as it appears if we investigate it by macroscopic, classical apparatuses.

Many years later, in 1948, the same way of reasoning was applied by Niels Bohr also to the language and logic that we use for the description of quantum phenomena.[19] In the meantime, 20 years had passed where various attempts and proposals for a new logic of quantum mechanics were published by a number of authors.[20] Bohr's critical reaction to these attempts refer in particular to the "three valued logic" proposed by Reichenbach.[21] Bohr's reaction to this and other attempts was:

> Incidentally, it would seem that the recourse to three-valued logic sometimes proposed as a means to dealing with the paradoxical features of quantum theory is not suited to give a clearer account of the situation, since all well-defined experimental evidence, even if it cannot be analysed in terms of classical physics, must be expressed in ordinary language making use of common logic.[22]

Taking together the old and the new arguments of Bohr, we find that his interpretation of quantum mechanics refers in a very consistent way to the classical world. Not only the measuring apparatuses are subject to classical physics, but also the concepts, the scientific language and its logic are taken from the classical world. Making use of this interpretation, the observer of the quantum world is strictly separated from this strange world and is thinking, speaking and operating exclusively in the familiar classical world.

[19] Bohr (1948).

[20] E.g. Birkhoff and von Neumann (1936) and Reichenbach (1944).

[21] Reichenbach (1944).

[22] Bohr (1948), p. 317.

5.4.2 The Contemporary Interpretation of Quantum Mechanics

This interpretation did not follow the attempts of the Copenhagen interpretation.[23] Instead, the scientific community made use of an alternative way, taking up a proposal that was indicated and partly elaborated already by J. von Neumann in his pioneering book[24] of 1932. Here we have in mind von Neumann's idea of applying quantum mechanics not only to microscopic quantum systems but also to the entire measurement process and its macroscopic apparatuses. Within the following five decades, it became more and more clear that the validity of quantum mechanics is not restricted to the microscopic world of molecules, atoms and elementary particles. Many macroscopic quantum effects such as superfluidity, superconductivity, and macroscopic tunnelling etc., confirmed that quantum mechanics can consistently be applied to these quantum phenomena. In this way it became obvious that quantum mechanics is indeed universally valid and applicable to all physical phenomena from atomic processes to the creation and evolution of the universe. In particular, quantum mechanics can be applied to the macroscopic apparatuses that are used for the experimental confirmation or refutation of several theoretical propositions. This means that quantum mechanics governs all processes that are needed for its own verification and falsification. In this way, classical apparatuses and classical processes are completely eliminated.

On the basis of these results, the quantum theory of measurements was developed by many authors in full detail. For more information we refer to the literature.[25,26] The result of these efforts can be summarised in terms of an improved new interpretation of quantum mechanics, which is based on the quantum theory of measurement and which is clearly distinguished from the original Copenhagen interpretation discussed above:

> Quantum mechanics describes the external, material reality as it appears if we investigate it by means of measuring apparatuses that are, on their part, physical objects and subject to the laws of quantum mechanics.[27]

This improved interpretation is not an approach exactly in the sense of realism but nearer to the quantum world than the old Copenhagen interpretation since it does not make use of an external classical semantics. However, it is still an "interpretation" that distinguishes clearly the theoretical terms from the elements of the physical reality. Also in this contemporary interpretation, quantum mechanics assumes a semantics in the sense of an instrumentalism.

---

[23] For all details, we refer to Jammer (1974).

[24] von Neumann (1932).

[25] Lahti and Mittelstaedt (1990), pp. 13–17.

[26] Busch et al. (1991, 1995, 1996) and Mittelstaedt (1998).

[27] Cf. Mittelstaedt (2011).

5.4.3 *Relaxation of the Classical Ontology O(C) – The Quantum Ontology*

We will continue now with option (B) mentioned above and take into account that the classical ontology $O(C)$ is loaded with hypotheses that cannot be justified by rational reasoning or by experimental evidence. Similar as in the case of special relativity, also in quantum mechanics there is a completely different way to understand this theory. Instead of removing the original contradictions between quantum mechanics and classical mechanics by an interpretation that refers to the apparatuses and the quantum theory of measurement, we can proceed in the following way: In a first step we have to make clear again that classical physics is neither intuitive nor plausible but loaded with hypotheses that can neither be justified by rational reasoning nor by experimental evidence. Hence, the alleged contradiction between quantum mechanics and our everyday experience, which is supposed to be governed by classical physics, becomes irrelevant. One component of this contradiction, the fictitious, classically determined everyday experience, cannot be justified.

However, the other component of the suspected contradiction, the quantum mechanics in Hilbert space, is very well justified. Indeed, quantum mechanics can be reconstructed on the same basis as classical mechanics if merely the insufficiently justified hypotheses are relaxed. In the present case of quantum mechanics, the two classical hypotheses $O(C)^1$ and $O(C)^2$, which are concerned with the structure of space and time and which had to be abandoned in the case of special relativity, can be preserved. However, the four other hypotheses $O(C)^3 - O(C)^6$ must be relaxed. In this way, we obtain the weaker hypotheses $O(Q)^3 - O(Q)^6$ of an adequate quantum ontology $O(Q)$. This quantum ontology $O(Q)$ reads:

- $O(Q)^3$ Any elementary property P can be tested with the result that either P or the counter property $\overline{P}$ pertains to the system.
- $O(Q)^4$ Objects are not "thoroughgoingly" determined.
- $O(Q)^5$ Hence for quantum objects there is no causality law.
- $O(Q)^6$ There is no conservation of substance for quantum objects.

Obviously, the weak hypotheses $O(Q)^3 - O(Q)^6$ follow from quantum mechanics and thus the ontology $O(Q)$ is *necessary* for this theory.

As shown in Chapter 3, and in particular in Section 3.5, on the basis of this relaxed quantum ontology $O(Q)$ quantum mechanics can be reconstructed by the following steps:[28]

1. We must establish a scientific language, i.e. a pragmatics, syntax, and semantics.
2. We have to establish a formal propositional language, and a formal propositional logic. i.e. in the present case we have to formulate the calculus of intuitionistic quantum logic.

[28] Mittelstaedt (2006).

3. The Lindenbaum-Tarski algebra of this calculus turns out to be an atomic, orthomodular lattice that fulfils the covering law.
4. Together with the mathematical results of Piron, Keller, and Solèr we finally arrive at a Hilbert space over the complex numbers.[29]

This reconstruction of quantum mechanics shows in detail that all the alleged "strange" and "surprising" results of quantum mechanics can be obtained in this way. On account of this reconstruction the quantum ontology *O*(*Q*) is also *sufficient* for quantum mechanics. Accordingly, nothing must be explained and made plausible by means of an "interpretation". On account of the rational reconstruction mentioned, quantum mechanics allows for a direct realistic understanding of its results. The well known, rather sophisticated interpretations of quantum mechanics are at bottom based on the same fundamental misunderstanding of classical physics as the interpretations of special relativity mentioned earlier.

5.5 Summary

For the three most important theories of modern physics, special relativity, general relativity, and quantum mechanics, we found the same situation with respect to the problem of interpretation. The strange and at first sight surprising results of the theories are not in conflict with our everyday experience, but they contradict classical mechanics and the underlying classical ontology *O*(*C*), which are erroneously often assumed to govern the laws of our everyday experience. These contradictions can, however, be removed by appropriate relaxations of the ontology *O*(*C*) which is loaded with hypotheses that can neither be justified by rational reasoning nor by experimental evidence. – The resulting weak ontologies *O*(*SR*), *O*(*GR*), and *O*(*Q*) are not only consequences of the theories of special relativity, general relativity, and quantum mechanics, respectively, and thus *necessary* for the theories *SR*, *GR*, and *Q*. In addition, the three ontologies *O*(*SR*), *O*(*GR*), and *O*(*Q*) are also *sufficient* for special relativity, general relativity, and quantum mechanics, since the theories *SR*, *GR*, and *Q* can be obtained by rational reconstructions on the basis of the ontologies *O*(*SR*), *O*(*GR*) and *O*(*Q*), respectively, and this is achieved without any additional hypotheses or empirical results.

With respect to the problem of interpretation these results imply, that additional interpretations of *SR*, *GR*, and *Q*, respectively, with a corresponding *classical-world semantics* are not needed at all. If we abandon or relax in each case those hypotheses of classical ontology that cannot be justified by rational reasoning or by experimental evidence, then the outcomes of *special relativity*, *general relativity* and *quantum mechanics* can be understood literally and without any additional interpretation. However, the approach we proposed in this chapter for avoiding

[29] Dalla Chiara et al. (2004), pp. 72–74.

contradictions, should not be misunderstood as a new competing interpretation of special relativity, the geometry of general relativity and quantum mechanics, respectively. Instead, it should show the reasons for our astonishment about the statements of modern physics and in this way also help to dissolve it. As we have seen, the reason for our astonishment is rooted in a misinterpretation of our ordinary experience, more precisely, in the unproved assumption that our ordinary experience is governed by classical mechanics. In other words, the goal of the present investigation is enlightenment and not the establishment of a new interpretation of the theories of modern physics. Indeed, we merely showed, how we could eliminate in principle the reasons for establishing interpretations of modern physics.

Chapter 6
Concluding Remarks

6.1 Intuitiveness and Truth in Physical Theories

In the beginning of the present treatise we discussed the observation that many scientists consider Newton's classical physics as understandable and intuitive, whereas modern physics of the twentieth century is judged as difficult to grasp and unintuitive. This assessment is shared by many physicists and presumably by the majority of philosophers of science. Here, we did not investigate the question why physicists as well as philosophers accept these statements as true – simply since we consider both theses as erroneous.

In Chapter 1 we tried to present an explanation of this conviction in two separate steps. First, we explained that Newton's classical physics is incomprehensible and unintuitive, since classical physics is loaded with hypotheses that originate from the metaphysics and theology of the seventeenth and eighteenth century. Since for these hypotheses there are no rational explanations and also no empirical evidence, they are indeed incomprehensible and unintuitive. Second, for similar and related reasons, the theories of modern physics are more comprehensible and more intuitive than Newton's classical physics, since these theories contain much less of the metaphysical hypotheses in question. These more general statements were explained in detail in the first chapter.

On the basis of these still very general results, we raised the intricate, more specific question: Is it possible to reconstruct the well known theories of modern physics merely by abandoning or relaxing some of the mentioned hypotheses of classical physics? Surprisingly, the answer to this question turned out to be positive in several cases. As a proof or at least as a justification of these theses, we reconstructed in the subsequent Chapters 2 and 3 the most important theories of modern physics, the theories of relativity and the non-relativistic quantum mechanics in Hilbert space.

The theory of Special Relativity is the most convincing example of our way of reasoning. We mention here the important result, shown in Section 1.3, that Newton's theory of space-time and classical mechanics is loaded with at least six

P. Mittelstaedt, *Rational Reconstructions of Modern Physics*,
Fundamental Theories of Physics 174, DOI 10.1007/978-94-007-5593-2_6,

ontological hypotheses $O(C)^1 \ldots O(C)^6$ which are neither intuitive nor justified by rational reasoning or by empirical evidence. By abandoning the hypothesis $O(C)^1$ of the existence of an absolute and universal time, we obtained the generalised Lorentz-transformation $T_{II'}(v,\omega)$, relativistic mechanics and the Minkowskian space-time $M(\omega)$, where $\omega = c$ is a property of the empty space-time. Hence, special relativity seems to be nothing but a less hypothetical version of classical mechanics. However, the following question remains also here: Is the theory of special relativity a priori true? Obviously not, since special relativity is still based on the other five hypotheses $O(C)^2 \ldots O(C)^6$ of classical mechanics, which were not abandoned or relaxed in our reconstruction. In addition, special relativity could depend on other hypotheses of classical mechanics that we are not aware of.

Even if we cannot assert that special relativity is a priori true, our way of reconstruction of this theory by eliminating one ontological hypothesis suggests that special relativity is at least closer to the empirical truth than classical mechanics and classical theory of space-time. It is, however, very hard to say what is precisely meant by the statement that a theory T_1 is closer to the truth than another theory T_2 if both theories T_1 and T_2 are not completely true but partly false.[1] Popper has tried to clarify this situation and to give a precise meaning to the phrase "closer to the truth". According to Popper,[2] a partly false theory T_1 is closer to the truth than another partly false theory T_2 if T_2 allows to derive more false statements than T_1. From a logical point of view, this definition might be a useful clarification of the phrase mentioned, but in the case of our comparison between classical mechanics and special relativity it is not directly applicable and helpful. Hence, we restrict the result of our consideration to the simple statement that special relativity is an improvement of classical mechanics, since it is based on fewer ontological hypotheses than classical mechanics. There is, however, one important remark that should be made here.

Compared with classical mechanics, special relativity is not based on new empirical results and not on new additional hypotheses. Hence, compared with classical mechanics, special relativity is an improvement. In addition, special relativity is also more "*intuitive*" than classical mechanics, if we distinguish here clearly the concepts of "*directly intuitive*" and "*indirectly intuitive*" as defined in section (1.1). Many of the new, and at first sight astonishing, features of special relativity are at closer inspection "indirectly intuitive" in the explained sense.

In the reconstruction of General Relativity we found many similarities with special relativity but also important differences. The starting point is again classical mechanics, which is loaded by the six ontological hypotheses $O(C)^1 \ldots O(C)^6$. Similarly to special relativity, general relativity is not a priori true, since in our reconstruction we abandoned only the two hypotheses $O(C)^1$ and $O(C)^2$. However, on the basis of this reduced ontology we could already reconstruct the pseudo-Riemannian structure of space-time, i.e. the pseudo-Riemannian 4-dimensional

[1] Weingartner (2000), chapter 9.

[2] Popper (1963, Appendix, 1972, p.330 ff).

space R_4. Obviously, this space-time model is closer to the empirical reality than the Minkowskian space-time of special relativity.

However, we should keep in mind that general relativity consists of two components, the Riemannian space-time R_4 and Einstein's field equations $G_{\mu\nu} = -\kappa T^{\mu\nu}$, where $G^{\mu\nu}$ is the Einstein tensor, $T^{\mu\nu}$ the energy-momentum tensor of matter and other fields and κ the gravitational constant. This second component is still missing in our reconstruction of general relativity. This means in particular, that we do not know anything about the coupling constant κ and above all, that we have no overview of those space-time models R_4 that are empirically meaningful. It is well known that the hypothetical assumption of Einstein's field equations implies – in particular in the large-scale region – many solutions that are far from being "directly intuitive" or "indirectly intuitive".[3]

The incorporation of Einstein's field equations into our reconstruction program of general relativity requires new hypothetical assumptions that are not justified by rational reasoning or by empirical evidence. Hence, we could generate in this way a situation in which the theory in question is again loaded with unjustified hypotheses, similar to Newton's classical mechanics. Of course, the two cases are quite different in detail. Whereas Newton's classical mechanics is loaded with metaphysical and theological hypotheses, the theory of general relativity is based on mathematical and formal hypotheses. For example, we mention here the requirements of *general covariance* and of *general relativity*, *Mach's principle*, and the requirement that the field equations should be the most simple quasi-linear second order differential equations.[4] For a present-day theoretical physicist, these hypotheses sound perhaps more convincing than the metaphysical hypotheses of Newton; they are, however, by no means really justified.

Whereas special relativity is a relaxed version of classical mechanics, which is based on less hypothetical assumptions and thus nearer to the empirical truth than classical mechanics, for general relativity this way of reasoning can only be applied to the first part of the theory that is concerned with the geometry of the pseudo-Riemannian space-time. The justification of the second part of general relativity, Einstein's field equations, is less convincing since this part is based on several mathematical as well as methodological hypotheses. Hence, also the implications of Einstein's field equations are not completely settled. Our method of reconstruction shows not only the advantages of the elimination of uncertain hypotheses but also the disadvantages of the addition of new hypothetical assumptions.

In the rational reconstruction of Quantum Mechanics we find quite different strategies. Since we reconstructed here only the non-relativistic quantum mechanics, the two hypotheses $O(C)^1$ and $O(C)^2$ were preserved. However, the abandonment of the remaining four hypotheses $O(C)^3$ … $O(C)^6$ leads to important restrictions of the possibilities of a formal scientific language, which can adequately

[3] Hawking and Ellis (1973).

[4] For more details cf. Wheeler (1973).

be expressed by a weakening of the formal logic of this language. We argue, that this reduced logic, the intuitionistic quantum logic, is a priori true in a much better justified sense than the pretended a priori truth of the well known classical logic.

It is obvious that these first results are very abstract and still far from a reconstruction of quantum mechanics. However, the intuitionistic quantum logic is the right first step, the beginning of a reconstruction of quantum mechanics. Indeed, after a long sequence of formal steps we arrive at the Lindenbaum-Tarski algebra of the calculus of quantum logic, at various lattice structures, and in particular at the lattice L_Q^* of the quantum logic of an individual quantum system. The goal of this approach is the reconstruction of quantum mechanics in Hilbert space *H(C)* over the complex numbers *C*. Up to this point no new empirical results were used and no additional formal assumptions were incorporated into our approach. Hence, the high degree of apriority of our results is still preserved.

At this point, the reconstruction must stop. The reason is that the lattice L_Q^* does not restrict the number fields of the Hilbert space in the desired way. There are not only three fields *R*, *C*, and *Q* – as originally expected according to the Piron-McLaren theorem – but infinitely many. Meanwhile we know a mathematical condition that excludes the superfluous number fields, the angle bisecting condition by Solèr, but this condition is a purely formal hypothesis. As long as this condition cannot be expressed and justified by arguments that refer to the most general possibilities of a scientific language of physics, the apriority of the reconstructed theory is not yet secured. If we were willing to accept the Solèr condition without mental reservation, then we could incorporate further refinements into the theory that refer to unsharp properties and to the uncertainty relation – and that without any loss of apriority. If the numerical value of Planck's constant $\hbar$ is accepted as an empirical component, then we could elaborate the modifications by unsharp properties even quantitatively. This is indicated in Section 4.3 and elaborated in detail in the literature.[5]

The gain of knowledge that we can receive by the method of rational reconstruction is obvious – without restriction – only in special relativity. In general relativity, we must assume new hypotheses that lead possibly also to unrealistic consequences. In contrast to quantum mechanics, where we could exclude unrealistic models by a convenient hypothesis, in general relativity a restricting condition of this kind is not known today. However, the assumption by means of which we could exclude unrealistic consequences in quantum mechanics – the angle bisecting condition – is presently still a useful hypothesis which can neither be proved by theoretical arguments nor by empirical evidence.

At the end of the long way to reconstruct the theories of modern physics by abandoning metaphysical hypotheses of Newton's classical world, there is still one open problem. In the real history of physics, the hypothetical components of classical physics were discovered by the founders of modern physics not at once, but step by step. We mention here the important contributions by *Enst Mach, Henri*

---

[5] Mittelstaedt (2008).

Poincaré, Albert Einstein, Niels Bohr, Werner Heisenberg, and Erwin Schrödinger and many others. On the basis of these discoveries, we can provide today rational reconstructions of the theories of modern physics. However, the pioneers of modern physics were only interested in those hypotheses of classical physics that are relevant for the construction of the new theories of the twentieth century. They did not investigate the problem whether there are other, not yet discovered, hypotheses of classical physics, which are still contained as hidden hypothetical components in the theories of modern physics.

The solution of this problem is not only interesting for the future development of physics, but also for the truth of presently accepted theories. If there were no residual presuppositions in these theories, according to the possible rational reconstruction, they could be considered as a priori true. This is, however very improbable, since all physical theories are based on several unavoidable metaphysical assumptions. We mention here for instance the assumption that there is an external world, outside of our consciousness. This hypothesis is well known and often called "metaphysical realism". Other options are the assumptions that the underlying ontology of physical theories is an ontology of objects and properties, as we have used it here, or an ontology of processes.[6] The decision between these options has presumably consequences for the language and logic which we use for the formulation of the physical theories. Many other, quite general assumptions of this kind are conceivable. We will not go into details here, but refer to the extended literature.[7] What counts here is that physics without any kind of metaphysics seems to be impossible. The elimination of metaphysical hypotheses of classical physics that we discussed in the present treatise refers only to avoidable hypotheses but not to metaphysical assumptions as a whole.

---

[6] Kuhlmann (2000, 2010).

[7] Vollmer (2000, 2007) and the literature quoted in these articles.

References

Armstrong, D. (1983). *What is a law of nature?* Cambridge: Cambridge University Press.

Becker, O. (Ed.). (1965). *Zur Geschichte der griechischen Mathematik.* Darmstadt: Wissenschaftliche Buchgesellschaft.

Birkhoff, G., & von Neumann, J. (1936). The logic of quantum mechanics. *Annals of Mathematics, 37*, 823–843.

Bohr, N. (1928). The quantum postulate and the recent development of atomic theory. *Nature, 121*, 580–590; Das Quantenpostulat und die neuere Entwicklung der Atomistik. *Die Naturwissenschaften, 16*, 245–257.

Bohr, N. (1934). *Atomic theory and the description of nature.* Cambridge: Cambridge University Press.

Bohr, N. (1948) On the notion of causality and complementarity. *Dialectica, 2*, 312–319.

Borchers, H. J., & Hegerfeldt, G. C. (1972). Über ein Problem der Relativitätstheorie: Wann sind Punktabbildungen des R^n linear? In *Nachrichten der Akademie der Wissenschaften in Göttingen, II.* Mathematisch-Physikalische Klasse, Nr.10 (pp. 205–229). Göttingen: Vandenhoeck & Ruprecht.

Born, M. (1920). *Die Relativitätstheorie Einsteins und ihre physikalischen Grundlagen (gemeinverständlich dargestellt).* Berlin: Springer.

Bose, S. N. (1924a). Plancks Gesetz und Lichtquantenhypothese. *Zeitschrift für Physik, 26*, 178–181.

Bose, S. N. (1924b). Wärmegleichgewicht und Strahlungsfeld in Abwesenheit der Materie. *Zeitschrift für Physik, 27*, 384–393.

Busch, P. (1985). Indeterminacy relations and simultaneous measurements in quantum theory. *International Journal of Theoretical Physics, 24*, 63–92.

Busch, P. (1998). Can 'unsharp objectification' solve the measurement problem? *International Journal of Theoretical Physics, 37*, 241–247.

Busch, P., & Shimony, A. (1996). Insolubility of the quantum measurement problem for unsharp observables. *Studies in History and Philosophy of Modern Physics, 27*, 397–404.

Busch, P., Grabowski, M., & Lahti, P. (1995). *Operational quantum physics.* Heidelberg: Springer.

Busch, P., Lahti, P., & Mittelstaedt, P. (1996). *The quantum theory of measurement* (2nd ed.). Heidelberg: Springer.

Busch, P., Heinonen, T., & Lahti, P. (2007). Heisenberg's uncertainty principle. *Physics Reports, 452*, 155–176.

Carr, B. (2007). *Universe or multiverse?* Cambridge: Cambridge University Press.

Castellani, E., & Mittelstaedt, P. (2000). Leibniz' principle, physics, and the language of physics. *Foundations of Physics, 30*, 1587–1604.

Curry, H. B. (1952). *Lecons de Logique Algébrique.* Paris: Gauthier – Villars.

P. Mittelstaedt, *Rational Reconstructions of Modern Physics*, Fundamental Theories of Physics 174, DOI 10.1007/978-94-007-5593-2,

Dalla Chiara, M. L. (1995). Unsharp quantum logics. *International Journal of Theoretical Physics, 34*, 1331–1336.

Dalla Chiara, M. L., & Giuntini, R. (1994). Partial unsharp quantum logics. *Foundation of Physics, 24*, 1161–1177.

Dalla Chiara, M. L., Cattaneo, G., & Giuntini, R. (1993). Fuzzy-intuitionistic quantum logic. *Studia Logica, 52*, 1–24.

Dalla Chiara, M. L., & Giuntini, R. (2001). Quantum logics. http://arxiv.org/pdf/quant-ph/0101028v2.pdf

Dalla Chiara, M. L., Giuntini, R., & Greechie, R. (2004). *Reasoning in quantum theory*. Dordrecht: Kluwer Academic Publishers.

Ehlers, J., Pirani, F., & Schild, A. (1972). The geometry of free fall and light propagation. In L. O'Raiffeartaigh (Ed.), *General relativity* (pp. 63–84). Oxford: Clarendon.

Einstein, A. (1917). *Über spezielle und allgemeine Relativitätstheorie (Gemeinverständlich)*. Braunschweig: Vieweg Verlag.

Einstein, A. (1924). *Quantentheorie des einatomigen idealen Gases*. Sitzungsberichte der Preussischen Akademie der Wissenschaften (Berlin), Physikalisch-mathematische Klasse (pp. 261–267).

Einstein, A. (1925). *Quantentheorie des einatomigen idealen Gases*. Zweite Abhandlung, Sitzungsberichte der Preussischen Akademie der Wissenschaften (Berlin), Physikalisch-mathematische Klasse (pp. 3–14).

Ellis, G. F. R. et al. (2003). Multiverse and physical cosmology. http://arxiv.org/pdf/astro-ph/0305292.pdf.

Falkenburg, B. (2006). Functions of intuition in quantum physics. In E. Carson & R. Huber (Eds.), *Intuition and the axiomatic method* (pp. 267–292). Dordrecht: Springer.

Feyerabend, P. K. (1965). Bemerkungen zur Verwendung nicht-klassischer Logiken in der Quantentheorie. In P. Weingartner (hrsg.), *Veröffentlichungen des internationalen Forschungszentrums für Grundlagen der Wissenschaften in Salzburg* (Band 1, pp. 351–359). Wien.

Feyerabend, P. K. (1970). Consolations for the spezialist. In I. Lakatos & A. Musgrave (Eds.), *Criticism and the growth of knowledge* (pp. 197–230). Cambridge: Cambridge University Press.

Foulis, D. J., & Bennett, M. K. (1994). Effect algebras and unsharp quantum logics. *Foundation of Physics, 24*, 1331–1352.

Franck, P., & Rothe, H. (1911). Über die Transformation der Raum-Zeitkoordinaten von ruhenden auf bewegte Systeme. *Annalen der Physik, 34*, 825–855.

Gamov, G. (1946). *Mr. Tompkins in wonderland*. New York: The Macmillan Company.

Giuntini, R. (1990). Brower-Zadeh logic and the operational approach to quantum mechanics. *Foundation of Physics, 20*, 701–714.

Giuntini, R., & Greuling, H. (1989). Towards a language of unsharp properties. *Foundation of Physics, 20*, 931–935.

Hawking, S. W., & Ellis, G. F. R. (1973). *The large scale structure of space-time*. Cambridge: Cambridge University Press.

Hentschel, K. (1990). *Interpretationen und Fehlinterpretationen der speziellen und allgemeinen Relativitätstheorie durch Zeitgenossen Albert Einsteins*. Basel: Birkhäuser.

Howard, D. (2004). Who invented the 'Copenhagen interpretation'? A study in mythology. *Philosophy of Science, 71*, 669–682.

Huber, R. (2000). *Einstein und Poincaré* (pp. 267–273). Paderborn: Mentis.

Huber, R. (2006). Intuitive cognition and the formation of theories. In E. Carson & R. Huber (Eds.), *Intuition and the axiomatic method* (pp. 293–324). Dordrecht: Springer.

Hund, F. (1967). *Geschichte der Quantentheorie*. Mannheim: Bibliographisches Institut.

Ignatowski v, W. (1910). Einige allgemeine Bemerkungen zum Relativitätsprinzip. *Physikalische Zeitschrift, 11*, 972–976.

Jammer, M. (1974). *The philosophy of quantum mechanics*. New York: Wiley.

Jauch, J. M. (1968). *Foundations of quantum mechanics*. Reading: Addison –Wesley.

Kalmbach, G. (1983). *Orthomodular Lattices*. London: Academic.

Kant, I. (1998). *The critique of pure reason* (Guyer, P., & Wood, A. Trans.). Cambridge: Cambridge University Press.

Keller, H. A. (1980). Ein nichtklassischer Hilbertraum. *Mathematische Zeitschrift, 172*, 41–49.

Könneker, C. (2001). *Auflösung der Natur– Auflösung der Geschichte*: moderner Roman und NS-"Weltanschauung" im Zeichen der theoretischen Physik. Stuttgart: Metzler.

Kuhlmann, M. (2000) Processes as objects of quantum field theory. In J. Faye, U. Scheffler, & M. Urchs (Hrg.), *Facts, things and events*: *Vol. 76 of Poznan studies* (pp. S. 365–388). Atlanta: Rodopi.

Kuhlmann, M. (2010). *The ultimate constituents of the material world*. Heusenstamm: Ontos-Verlag.

Kuhn, T. S. (1962). *The structure of scientific revolutions* (1st ed., 1970 2nd ed.). Chicago: University of Chicago Press.

Lahti, P., & Mittelstaedt, P. (Eds.). (1990). *Symposium on the foundations of modern physics 1990 – Quantum theory of measurement and related philosophical problems, Joensuu, Finland, 13–17 August 1990*. Singapore: World Scientific.

Laugwitz, D. (1960). *Differentialgeometrie*. Stuttgart: B. G. Teubner.

Leibniz, G. W. Philosophische Schriften (GP). C. I. Gerhardt (Ed.). Berlin 1875–1890.

Lévy-Leblond, J. M. (1976). One more derivation of the Lorentz transformation. *American Journal of Physics, 44*, 271–277.

Linde, A. D. (1990). *Inflation and quantum cosmology*. San Diego/London: Academic Press, Inc.

Lorenzen, P. (1955). *Einführung in die operative Logik und Mathematik*. Heidelberg: Springer.

Mach, E. (1901). *Die Mechanik in ihrer Entwicklung* (4th ed.). English translation: Brockhaus, F. A., Leipzig. (1902) *The science of mechanics* (trans: McCormack, T.J.). Chicago.

MacLaren, M. D. (1965). Nearly modular orthocomplemented lattices. *Transactions of the American Mathematical Society, 114*, 401–416.

Martzke, R., & Wheeler, J. A. (1964). Gravitation as geometry I: The geometry of space-time and the geometrodynamical standard meter. In H. Y. Chiu & W. F. Hoffman (Eds.), *Gravitation and relativity* (pp. 40–64). New York: A. Benjamin.

Misner, C. W., Thorne, K. S., & Wheeler, J. A. (1973). *Gravitation*. San Francisco: W. H. Freeman.

Mittelstaedt, P. (1976a). *Philosophical problems of modern physics*. Dordrecht: D. Reidel Publishing Company.

Mittelstaedt, P. (1976b). *Der Zeitbegriff in der Physik*. Mannheim: BI-Wissenschaftsverlag.

Mittelstaedt, P. (1978). *Quantum logic*. Dordrecht: D. Reidel Publishing Company.

Mittelstaedt, P. (1995). *Klassische Mechanik* (2nd ed.). Mannheim: Bilbliographisches Institut.

Mittelstaedt, P. (1998). *The interpretation of quantum mechanics and the measurement process*. Cambridge: Cambridge University Press.

Mittelstaedt, P. (2003). Hätte Newton die Relativitätstheorie finden können? In W. Buschlinger & C. Lütge (Eds.), *Kaltblütig: Philosophie von einem rationalen Standpunkt* (pp. 221–240). Stuttgart: S. Hirzel Verlag.

Mittelstaedt, P. (2005). Quantum physics and classical physics – In the light of quantum logic. *International Journal of Theoretical Physics, 44*, 771–781.

Mittelstaedt, P. (2006). Intuitiveness and truth in modern physics. In E. Carson & R. Huber (Eds.), *Intuition and the axiomatic method* (pp. 251–266). Dordrecht: Springer.

Mittelstaedt, P. (2008). Planck's constant in the light of quantum logic. *International Journal of Theoretical Physics, 47*, 104–113.

Mittelstaedt, P. (2011). The problem of interpretation of modern physics. *Foundations of Physics, 41*, 1667–1676.

Mittelstaedt, P., & Weingartner, P. (2004). *Laws of nature*. Heidelberg: Springer.

Mittelstaedt, P., Prieur, A., & Schieder, R. (1987). Unsharp particle-wave duality in a photon split-beam experiment. *Foundations of Physics, 17*, 891–903.

Natanson, L. (1911). Über die statistische Theorie der Strahlung. *Physikalische Zeitschrift, 12*, 659–666.

Newton, I. (1687). *Philosophia Naturalis Principia Mathematica*. London: Streator.

Newton, I. (1934). *Mathematical principles of natural philosophy*. Trans. A. Motte (1729), rev. F. Cajori. Berkeley: University of California Press.

Pauli, W. (1940). On the connection between spin and statistics. *Physics Review, 58*, 716–722.

Penrose, R. (1959). The apparent shape of a relativistically moving sphere. *Mathematical Proceedings of the Cambridge Philosophical Society, 55*, 137–139.

Petrow, A. S. (1964). *Einstein-Räume*. Berlin: Akademie Verlag.

Piron, C. (1964). Axiomatique Quantique. *Helveti Physics Acta, 37*, 439–468.

Piron, C. (1976). *Foundations of quantum physics*. Reading: W. A. Benjamin.

Poincaré, H. (1898). La mesure du temps. *Revue de metaphysique et de moral, VI*, 1–13.

Popper, K. (1963). *Conjectures and refutations*. London: Routledge & Keagan Paul.

Popper, K. (1972). *Objective knowledge: An evolutionary approach*. Oxford: Clarendon Press.

Putnam, H. (1969). Is logic empirical? In M. Wartofsky & R. Cohen (Eds.), *Boston studies in the philosophy of science* (Vol. 5). New York: Humanities Press.

Reichenbach, H. (1944). *Philosophical foundation of quantum mechanics*. Berkeley/Los Angeles: University of California Press; Deutsche Übers. (1949). *Philosophische Grundlagen der Quantenmechanik*. Basel: Birkhäuser.

Sakharov, A. D. (1967). Vacuum quantum fluctuations in curved space and the theory of gravitation. *Doklady Akad. Nauk S.S.S.R., 177*, 70–71. English translation *Soviet Physics Doklady 12*, 1040–1041.

Schröter, J., & Schelb, U. (1994). A space-time theory with rigorous axiomatics. In U. Majer & H.-J. Schmidt (Eds.), *Semantical aspects of spacetime theories*. Mannheim: BI-Wissenschaftsverlag.

Sexl, R. U., & Urbantke, H. K. (1992). *Relativität, Gruppen, Teilchen* (3rd ed.). New York: Springer.

Solèr, M.-P. (1995). Characterisation of Hilbert spaces by orthomodular lattices. *Communications in Algebra, 23*(1), 219–243.

Stachel, J. (1982). Einstein and Michelson: The context of discovery and the context of justification. *Astronomische Nachrichten, 303*, 47–53.

Stachel, J. (1989). Quale cancone cantarono le sirene. In U. Curi (Ed.), *L'opera di Einstein* (pp. 21–37). Ferrara: Gabriele Corbino & Company.

Stachel, J. (2002). *Einstein from B to Z*. Boston: Birkhäuser.

Stachow, E. W. (1976). Completeness of quantum logic. *Journal of Philosophical Logic, 5*, 237–280.

Stachow, E. W. (1980). Logical foundations of quantum mechanics. *International Journal of Theoretical Physics, 19*, 251–304.

Stachow, E. W. (1984). Structures of quantum language for individual systems. In P. Mittelstaedt & E.-W. Stachow (Eds.), *Recent developments in quantum logic* (pp. 129–145). Mannheim: BI-Wissenschaftsverlag.

Stegmüller, W. (1970). *Probleme und Resultate der Wissenschaftstheorie und Analytischen Philosophie: Vol. 2. Theorie und Erfahrung*. Berlin/Heidelberg: Springer.

Streater, R. F., & Wightman, A. S. (1964). *PCP, spin and statistics and all that (2000)* (2nd ed.). Princeton: Princeton University Press.

Terrell, J. (1959). Invisibility of the Lorentz contraction. *Physical Review, 116*, 1041–1045.

Van Fraassen, B. (1989). *Laws and symmetries*. Oxford: Oxford University Press.

Varadarajan, V. S. (1968). *Geometry of quantum theory* (Vol. 1). Princeton: Van Nostrand.

Vilenkin, A. (1982). Creation of universes from nothing. *Physics Letters, 117 B*, 25–28.

Vollmer, G. (2000). Was ist naturalismus? In G. Keil & H. Schnädelbach (Eds.), *Naturalismus* (pp. 46–67). Frankfurt a. Main: Suhrkamp.

Vollmer, G. (2007). Wie viel Metaphysik brauchen wir? In D. Westerkamp & A. Lühe (Eds.), *Metaphysik und Moderne* (pp. 67–81). Würzburg: Königshausen & Neumann.

von Helmholtz, H. (1868). *Über die Tatsachen, die der Geometrie zu Grunde liegen.* Nachrichten der Königlichen Gesellschaft der Wissenschaften und der Georg-Augusts-Universität, Nr. 9 (pp. 193–221).
von Neumann, J. (1932). *Mathematische Grundlagen der Quantenmechanik.* Berlin: Springer.
Weingartner, P. (2000). *Basic questions on truth.* Dordrecht: Kluwer Academic Publishers.
Wheeler, J. A. (1973). From relativity to mutability. In J. Mehra (Ed.), *The physicist's conception of nature* (pp. 202–247). Dordrecht: Reidel.

Index

P. Mittelstaedt, *Rational Reconstructions of Modern Physics*, Fundamental Theories of Physics 174, DOI 10.1007/978-94-007-5593-2,

MIX
Papier aus verantwortungsvollen Quellen
Paper from responsible sources
FSC® C105338

If you have any concerns about our products,
you can contact us on
ProductSafety@springernature.com

In case Publisher is established outside the EU,
the EU authorized representative is:
Springer Nature Customer Service Center GmbH
Europaplatz 3, 69115 Heidelberg, Germany

Printed by Libri Plureos GmbH
in Hamburg, Germany